INDIAN PHYCOLOGICAL REVIEW

Vol. 2 1993

SPECIAL SERIES - I.

CYANOPHYTA
(CYANOBACTERIA)

Edited
by
N. Anand
C.A.S. in Botany
University of Madras
Madras 600 025

M/S Bishen Singh Mahendra Pal Singh
P.O. Box No. 137, Dehra Dun - 248 001
[INDIA]

INDIAN
PHYCOLOGICAL REVIEW VOL. 2

ISBN 81-211-0083-6

Printed by Gajendra Singh Gahlot at Shiva Offset Press,
and composed by Shivalik Graphic Services
14, Old Connaught Place, Dehra Dun, India
for Bishen Singh Mahendra Pal Singh, Dehra Dun, India.

INDIAN PHYCOLOGICAL REVIEW

LIST OF CONTRIBUTORS

1. Dr A.K. Ahluwalia,
   Dept. Botany, Panjab Uni.,
   Chandigarh 160 014.

2. Dr N. Anand,
   CAS in Botany, Uni. Madras, Guindy,
   Madras 600 025

3. Dr S.G. Bharathi,
   Dept. Botany, Karnataka Uni.,
   Dharwad 580 003

4. Dr S. Kannaiyan,
   Dept. Agricultural Microbiology,
   Tamil Nadu Agricultural Uni.,
   Coimbatore 641 003

5. Dr A.K. Kashyap,
   CAS in Botany, Banaras Hindu Uni.,
   Vananasi 221 005

6. Dr B.D. Kaushik,
   National Facility of blue-green algae,
   I.A.R.I., New Delhi 110 012

7. Dr P. Prasad,
   CAS in Botany, Banaras Hindu Uni.,
   Varanasi 221 005

8. Dr G. Revathy,
   CAS in Botany, Uni. Madras,
   Madras 600 025.

9. Dr V. Sankaran,
   Dept. Bot., N.G.M. College,
   Pollachi 642 001

10. Dr P.K. Singh,
    Central Rice Research Institute,
    Cuttack 753 006

11. Dr S.P. Singh,
    CAS in Botany, Banaras Hindu Uni.,
    Varanasi 221 005.

DEDICATED
TO

My esteemed Teacher

Prof. T.V. Desikachary., D.Sc., F.N.I., F.A. Sc.,

CONTENTS

Topic			Pages
1.	Foreword		
2.	Preface		
3.	Preamble	::	1
4.	Morphology, classification and taxonomic studies N. Anand	::	3
5.	Biosystematics V. Sankaran	::	19
6.	Reproduction, differentiation and developmental cycles N. Anand & G. Revathy	::	31
7.	Habitats and distribution A.S. Ahluwalia	::	41
8.	Heavy Metal Biology S.P. Singh	::	59
9.	Technology for rice crop S. Kanniyan	::	73
10.	Biology of cyanophages A.K. Kashyap, P. Prasad & P.K. Singh	::	83
11.	Methods of study B.D. Kaushik	::	113
12.	Phylogeny and interrelationships S.G. Bharathi	::	139

APPENDIX : Research laboratories in India working on blue greens. :: 153

FOREWORD

The last few decades have seen the cyanobacteria attracting serious attention all over the world in studies relating to physiology, ecology, genetics, biotechnology and aquaculture. Extensive contributions on diverse aspects of the biology of cyanobacteria have emanated from India, notably from the Banaras and the Madras Schools. In some measure, the more modern and fashionable molecular biological aspects have overshadowed the more fundamental but no less important taxonomical and morphological issues. Happily, however, this unfortunate trend is showing some signs of reversal in the light of the recent surge of global interest in biodiversity, and the ascent of biodiversity to the top of the environmental agenda in many countries. The importance of collection, identification and cataloguing of cyanobacterial diversity from a wide variety of benign as well as harsh environments in the tropical and subtropical areas cannot be overemphasized. In this context, the present review edited by Dr. N. Anand is a timely addition to phycological knowledge and should help attract discerning young researchers to enter and enjoy the cyanobacterial world.

April 15, 1993.

H.D. Kumar
DEAN
FACULTY OF SCIENCE
BANARAS HINDU UNIVERSITY

PREFACE

"The honour is a pleasure but the responsibility is onerous" this was my feeling while acceding to edit this volume on Cyanophyta (Cyanobacteria). The ordeal of reviewing the information on different aspects of the Blue-greens demands a concerted effort of scientists working on diverse areas and I am absolutely grateful to all my fellow contributors to this volume. Their unfailing cooperation was a major source of inspiration to launch on this effort and fulfill my avocation of completing the task. I am delighted to have been able to do it. It has been a sincere attempt to cover several topics concerning the Blue-greens, but yet, certain aspects had to be left out judiciously, to be accommodated in the other edited volumes of the series. I would like to confess that due to paucity of information from Indian work, work done from elsewhere too had to be included in some chapters. Another point I would like to emphasize is that the Blue-greens are dealt as a very important group of plants of evolutionary significance, hence, setting aside all controversies of nomenclature the authors were given the option to name the Blue-greens - as they belived.

I remain indebted to Prof. T.S. Sadasivan, Prof. C.V. Subramanian, former Directors, CAS in Botany and Prof. A. Mahadevan, Director, CAS in Botany, University of Madras for their encouragements. I sincerely thank Prof. W.D.P. Stewart, D.Sc., F.R.S., F.R.S.E., University of Dundee, Dundee U.K. for his invaluable advices. My gratefulness is due to all fellow phycologists for sending their publications and all my students for their inquisitiveness. My thanks are due to Prof. V.N. Raja Rao for useful suggestions and Dr. R. Rengasamy for his help. Finally, I thank the architect of this endeavour the late Dr Mahmood Khan, Managing Editor of this series who entrusted this exciting effort with singular purpose - "prepare a poignant document for posterity".

February, 1993.

N. Anand

Ind. Phyco. Rev. 1993.
2: 1-2.

BLUE-GREEN ALGAE (CYANOPHYTA/CYANOBACTERIA) - A PREAMBLE

N. ANAND.
Centre of Advanced Study in Botany,
University of Madras, Madras - 600 025.

Blue-green algae are autotrophic prokaryotic organisms representing one of the earliest of living system, predominant even in the precambrian era. The structure of blue-green algal cell resembles the bacterial cell, but photosynthesis, the primary metabolic process is exactly like the eukaryotic algae and higher plants. Fogg *et. al.* (1973) rightly consider that these organisms "....... represent a pervading and vital component of the evolutionary tree which has made possible the highest forms of plant life". The nomenclature of blue-greens was a point of discussion with claims from phycologists to retain status within algae as Cyanophyta (Anand, 1981) and counter claims to rename them as Cyanobacteria (Stanier *et. al.* 1978). It seems the view of Carr and Whitton (1982) that '........it is likely that both codes will continue to be used in parallel for some more years to come.....' will hold good for a longer time. Lest the nomenclature, the blue-greens both structurally and functionally comprise of a very unique group of organisms known to biologists.

The occurrence of blue-green algae in cultivated paddy fields in India was first reported in 1939 by De and this finding incited the attention of several scientists all over the world. Studies in the last 3-4 decades on this group of organisms has been not only to understand their role in the conservation of nitrogen but also in the understanding of the problem of nitrogen metabolism. Blue-greens were model organisms to understand several metabolic processes, as several processes are combined astutely within a single cell. The process of nitrogen fixation, a higly oxygen sensitive reaction, can take place in the same cells that release oxygen during photosynthesis. The two processes go on unabated. Several blue-green can live heterotrophical, though basically all are autotrophic. There are examples of species that can switch over to anoxygenic photosynthesis from oxygenic photosynthesis. Ecologically, these are the most adaptive organisms. They occur virtually in all extremes of physically and nutritionally diverse habitats, both aquatic and terrestrial ecosytsems besides being present as airspores. In the aquatic environ they can be found as thermophilic and cryophilic forms; fresh water and saline forms; alkalotolenat and acidophilic forms. Among the terrestrial forms these can be found in all types of soils (both desert and fertile) and even in subaerial situations. Another unique feature of this group is that this is the only group that has representative members found in symbiotic association with all other plant types known besides a few lower animals (Fogg *et. al.* 1973).

In fact, the blue-green algal chloroplast itself is considered by phylogenetists to have played a role in the evolution of eukaryotic cells.

Considerable amount of research has been carried out in understanding the biology of blue-green algae in our country and elsewhere. The taxonomy of these organisms whether as Cyanophyceae under algae or as Cyanobacteria under bacteria has always been not without problems. Stable characters whether morphological or physiological or biochemical are still to be evolved to identify the taxa at species level. In fact, the natural material under culture may bring into question the very recognition of several species (Anand, 1988). Physiologically also the behaviour of different strains identified as same species do not always remain similar. One has to search for strong physiological character to delineate the species.

Experiments on the genetic transformation in blue-greens generated enormous amount of information especially in differentiation of various cell types, nitrogen fixation and resistance to drugs in the last two decades. However, evidences of genetic transformation at molecular level seems to be a requirement to substantiate all earlier studies. The use of blue-greens as biofertilizers to conserve the chemical fertilizer input has been an area of great utility to the farming community of India. A conservation of nearly 30% of chemical fertilizer input has been demonstrated when blue-greens are used as co-fertilizers. *Spirulina* has been identified as a good source of single cell protein. Blue-greens are used in waste water treatment and recent attempts are to use them as a non-conventional source of energy. Several strains produce hydrogen which may be a substitute fuel in future.

REFERENCES

Anand, N. 1981. Certain viewpoints on renaming Cyanophyta. *Curr. Sci.* **50:** 535-536.

Anand, N. 1988. Culture studies and taxonomy of blue-green algae-certain identification problems. *Arch. Hydrobiol. Suppl.* **80:** 141-147.

Carr, N.G. and Whitton, B.A. 1982. *The Biology of Cyanobacteria.* Blackwell Scientific Publications, London.

Fogg, G.E., Stewart, W.D.P., Fay, P., and Walsby, A.E., 1973. *The Blue-green Algae.* Academic Press, London.

Stanier, R.Y., Sistrom, W.R., Hansen, T.A., Whitton, B.A., Castenholz., R.W., Pfennig, N., Gorlenko, V.N., Kondratieva, E.N., Einhjellen, R., Whittenbury,R., Gherna, R.L. and Trueper, H.G. 1978. Proposal to replace the nomenclature of the Cyanobacteria (blue-green algae) under the nomenclature of bacteria. *Int. J. Syst. Bactkeriol.* **28:** 335-336.

Ind. Phyco. Rev. 1993.
2: 3-18.

MORPHOLOGY, CLASSIFICATION AND TAXONOMIC STUDIES

N. ANAND,
CAS in Botany, University of Madras,
Guindy, Madras - 600 025.

Blue-green algae (Cyanophyta) are prokaryotic organisms with oxygen evolving photosynthetic process. Their occurrence in the sediments from early precambrian period suggests that these microorganisms could have played a pivotal role in the evolution of the aerobic environment. The major pigments in blue-green algae are chlorophyll and the water soluble biliproteins the phycobilins, comprising of phycocyanin, phycoerythrin and allophycocyanin. Phycobilins serve as the primary light harvesting pigments for photosystem II in these organisms. They lack flagella but some of them show a characteristic gliding movement. Sexual reproduction is absent and food reserves are glycogen and cyanophycin (Fogg *et al.* 1973).

Blue-green algae comprise of microscopic organisms that could be included under two broad categories of morphological organization namely, unicellular or coccoid forms and filamentous forms. A considerable degree of modification and diversification of the two basic types of morphological forms, results in a variety of thallus that ranges from a simple unicellular organism to a branched filamentous member.

The coccoid members consist of simple unicells that may be ovoid, spherical or cylindrical occurring as free cells or cells aggregated into a mucilaginous matrix. Some unicellular forms have firm sheath for each individual cells that group together to form a colony, besides the colonial sheath. Ordered arrangements of cells in a definite shaped colony (plate like or cubical) and irregularly arranged colonies form another set of coccoid members. Polarity of the cells, production of exospores or endospores (now called baeocytes), pseudoparenchymatous and pseudofilamentous nature of the thallus differentiate yet another group of coccoid blue-green algae.

The filamentous forms are separated on the basis of the presence or absence of specially differentiated cells called heterocysts into non-heterocystous or heterocystous filaments forms. Heterocystous filamentous form may be of uniform width or showing base apex differentiation, and there are forms that exhibit various types of true branching and false branching. Those exhibiting true branching are the heterotrichous forms with main and lateral branches, the highest level of differentiation seen in blue-green algae.

All characters mentioned above form the basis of the classification of blue-green algae and grouping of the organisms under various orders according to Geitler (1932) and Desikachary (1959). Cyanophyta is dealt as separate families without orders by Drouet (1951). Later revision by Drouet & Daily (1956) on coccoid forms and Drouet

(1967, 1972, 1978) on non-heterocystous and heterocystous filamentous forms were based on the consideration of ecological variablility. The revision which reduced the taxa of blue-green to an alarming low number is not generally accepted by most of the blue-green algal taxonomists. Recently, blue-green algae are included under Bacteria and termed Cyanobacteria (Bergey's manual) mainly based on their similarity to bacteria in the prokaryotic organization of the cell structure as revealed by electrol microscopic investigation. Nevertheless even in the classification of cyanobacteria into various typological groups the major criteria used have been the morphological diversity of the organisms (Rippka *et al.* 1979). A resume of the classification of blue-green algae by Desikachary (1959) and Rippka *et al.* (1979) is given below.

Summary of the classification of Cyanophyta by Desikachary (1959) and Rippka *et al.* (1979) renaming the same organisms as Cyanobacteria (after Anand, 1990).

Morphology	Order (Desikachary, 1959)	Section (Rippka *et al.* 1979)
Unicellular or colonial forms, dividing in 1, 2 or more planes	Chroococcales	I
Unicellular or colonial forms attached	Chamaesiphonales	II
Pseudoparenchymatous	Pleurocapsales	II
Filamentous trichomes without and differentiation	Nostocales (Oscillatoriaceae)	III
Filamentous with heterocysts and akinete differentiation	Nostocaceae	IV
Filamentous showing base and apex differentiation	Rivulariaceae Microchaetaceae	IV
Filamentous showing false branching	Scytonemataceae	IV
Filamentous showing main types of true branching like true lateral branching, reverse v shaped branching	Stigonematales	V

Diagnostic keys to the orders and some of the genera of blue-green algae, illustrations to genera are given in figures 1 - 46.

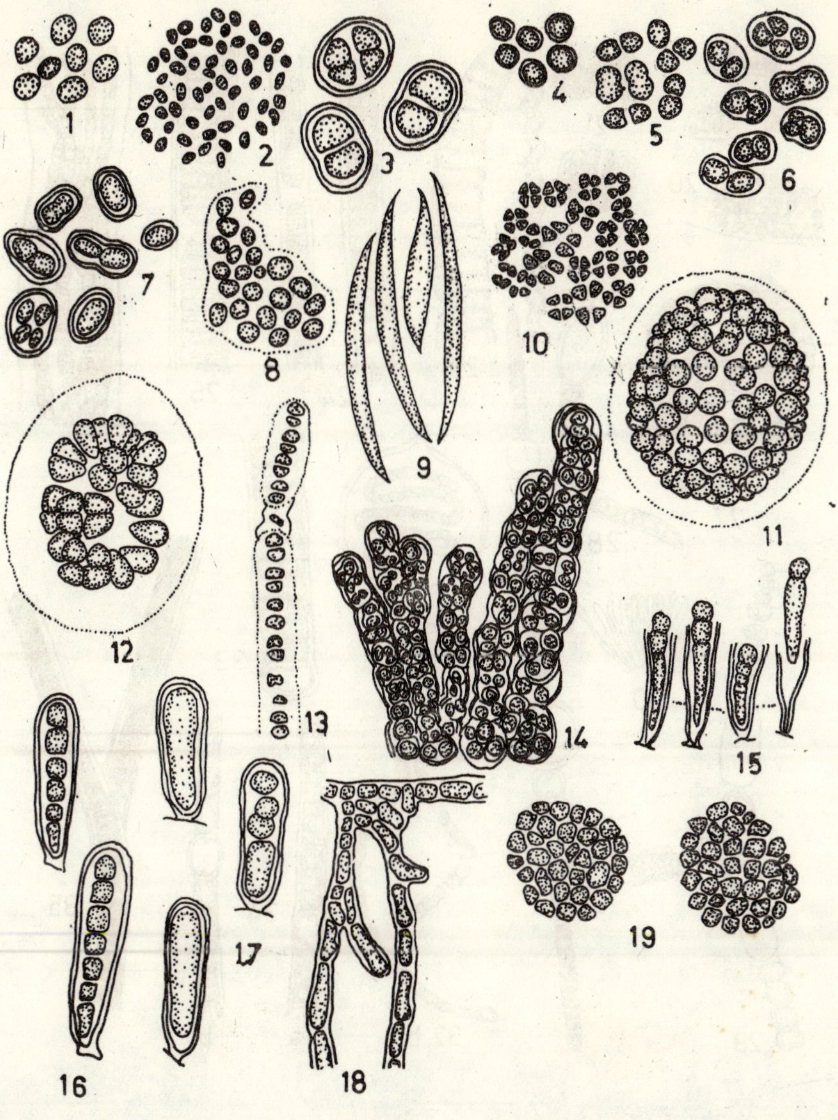

Figures 1-19. 1-10 redrawn from Anand (1990); Fig. 11-19 from Desikachary (1959).

1. *Synechocystis crassa* Woronich. 2. *Synechococcus elongatus* Naeg. 3. *Chroococcus tenax* (Kirchn.) Hiron. 4. *Aphanocapsa banaresensis* Bharadwaja 5. *Aphanothece castagnei* (Breb.) Rabenh. 6. *Gloeocapsa nigrescens* Naeg. 7. *Gloeothece rupestris* (Lyngb.) Bornet. 8. *Microcystis aeruginosa* Kuetz. 9. *Dactylococcopsis raphidioides* Hansg. 10. *Merismopedia glauca* (Ehr.) Naeg. 11. *Coelosphaerium dubium* Grunow. 12. *Gomphosphaeria aponina* Kuetz. 13. *Johannesbaptistia pellucida* (Dickie) Taylor et Drouet. 14. *Entophysalis granulosa* Kuetz. 15. *Chamaesiphon rostafinskii* (Rostaf.) Hansg. 16. *Stichosiphon sansibaricus* (Hieron.) Drouet & Daily. 17. *Dermocarpa clavata* Geitler. 18. *Hyella caespitosa* Born. et Flah. 19. *Hydrococcus rivularis* Kuetz.

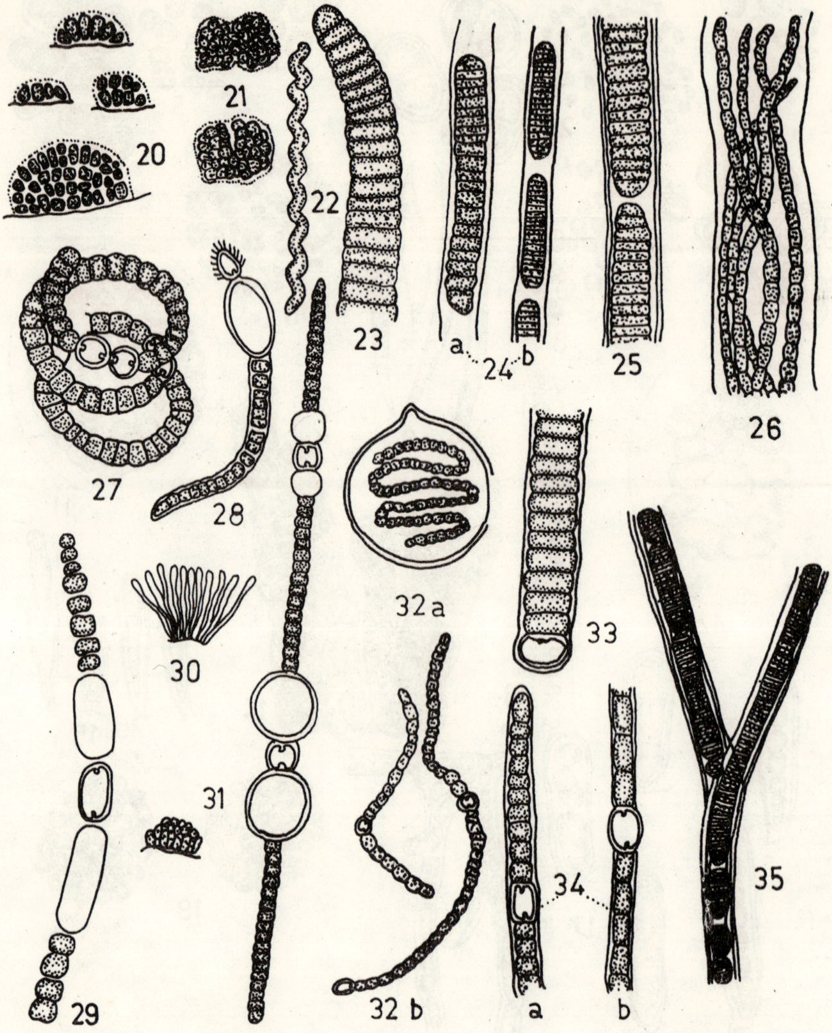

Figures 20-35. Figs. 22-26,28,29,32,34 redrawn from Anand (1990); Figs. 20,21,27,30,31,33, and 35 from Desikachary (1959).

20. *Xenococcus kerneri* Hansg. 21. *Myxosarcina burmensis* Skuja. 22. *Spirulina princeps* West, W. and G.S. 23. *Oscillatoria nigroviridis* Thwaites ex Gomont. 24a. *Phormidium ambiguum* Gomont 24b. *P. anomala* Rao, C.B. 25. *Lyngbya confervoides* Ag. 26. *Microcoleus acutissimus* Gardner. 27. *Anabaenopsis arnoldii v. indica* Ramanathan. 28. *Cylindrospermum stagnale* (Kuetz) Born. et Flah. 29. *Anabaena oscillarioides* Bory. 30,31. *Wollea bharadwajae* Singh, R.N. 32a. *Nostoc commune* Vaucher ex Born. et Flah. 32b. *N. muscorum* Ag. ex Born. et Flah. 33. *Nodularia spumigena* Mertens. 34a. *Aulosira fertilissima* Ghose. 34b. *A. prolifica* Bharadw. 35. *Plectonema wollei* Farlow.

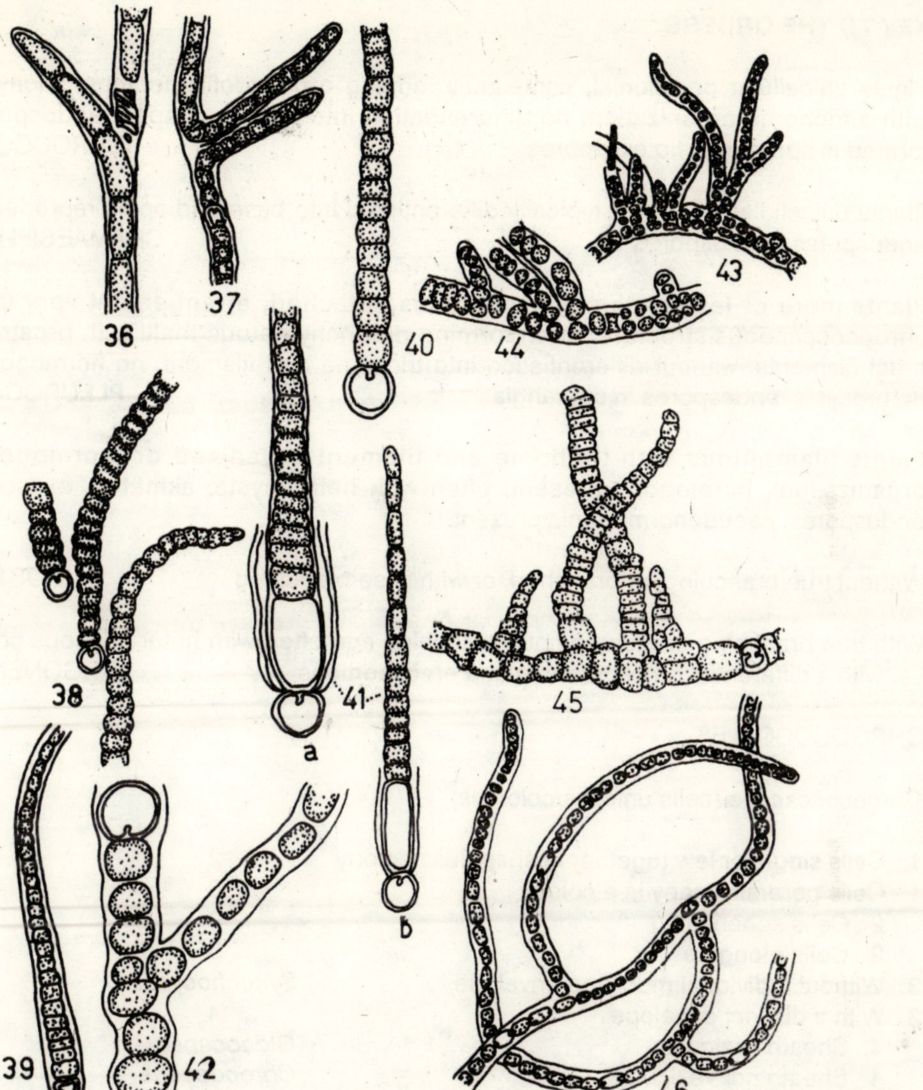

Figures 36-46. Figs. 37-42 and 45 redrawn from Anand (1990); Figs. 36,43,44 and 46 from Desikachary (1959).

36. *Scytonematopsis kashyapi* (Bharadw.) Geitler. 37. *Scytonema bohneri* Schmidle. 38. *Tolypothrix byssoidea* (Berk.) Kirchner. 39. *Calothrix castelli* (Massal.) Born. et Flah. 40. *Rivularia aquatica* De Wilde. 41a. *Gloeotrichia ghosei* Singh, R.N. 41b. *G. intermedia* (Lemm.) Geitler. 42. *Mastigocladus laminosus* Cohn. 43. *Stigonema dendroidium* Fremy. 44. *Fischerella epiphytica* Ghose. 45. *Westiellopsis prolifica* Janet. 46. *Hapalosiphon intricatus* West & West.

KEY TO THE ORDERS

Plants unicellular or colonial, sometimes forming a pseudofilamentous colony, never with a trichome organization, no differentiation into base and apex, endospores not formed in sporangia; no exospores CHROOCOCCALES

Plants unicellular, attached, typically differentiated into base and apex, reproduction by endospores or exospores CHAMAESIPHONALES

Plants more or less distinctly filamentous, attached, arrangement very uniform, chroococcaceous structure, often forming parenchymatous thalli with prostrate and erect filaments, without differentiation into trichome and filament, no hormogones, no heterocysts, endospores in sporangia PLEUROCAPSALES

Plants filamentous with trichome and filament organized or 'hormogonalean organization', hormogons present, often with heterocysts, akinetes, exospores or endospores, pseudohormogonia present.

Without true branching, unbranched, or with false branching NOSTOCALES

With true branching or dichotomous branching and often with heterotrichous condition, i.e. with a differentiation of prostrate and erect portions STIGONEMATALES

CHROOCOCCALES

Chroococcaceae (cells unicells/colonies)

1	Cells single or few together in shapeless colony	... 2
1	Cells gererally many in a colony	... 6
2	Cells spherical	... 3
2	Cells elongate	... 5
3	Without individual mucilage envelope	*Synechocystis*
3	With a distinct envelope	... 4
4	Sheath vesicular	*Gloeocapsa*
4	Sheath not vesicular	*Chroococcus*
5	Cell division with firm vesicular sheath	*Gloeotheca*
5	Sheath absent, cells 2-4 together in common mucilage	*Synechococcus*
6	Cells with definite arrangement	... 9
6	Cells without definite arrangement	... 7
7	Cells typically well packed into microscopic colonies, definite shaped planktonic	*Microcystis*
7	Cells loosely arranged microscopic colonies	... 8
8	Cells spherical	*Aphanocapsa*
8	Cells cylindrical	*Aphanothece*
8	Cells with tapering ends forming spike shaped colonies	*Dactylococcopsis*

9 Cells tubular or flat colonies — *Merismopedia*
9 Cells cubically arranged — *Eucapsis*
9 Cells arranged in hollow spherical — ... 10
10 Mucilage homogenous, cells spherical — *Coelosphaerium*
10 Mucilage not homogenous, cells pearshaped — *Gomphosphaeria*

Entophysalidaceae (cells arranged as pseudofilaments)

1 Cells in single row formig pseudo-filamentous thallus — *Johannesbaptistia*
1 Cells in more than single row — ... 2
2 Cells in distinct row, thallus crustose — *Entophysalis*
2 Thallus gelatinous and hollow — *Placoma*

CHAMAESIPHONALES

1 Attached thallus producing exospores (Chamaesiphonaceae) — *Chamaesiphon*
1 Attached thallus producing endospores (Dermocarpaceae) — ... 2
2 Endospores produced in one direction — *Stichosiphon*
2 Endospores produced in three directions — *Dermocarpa*

PLEUROCAPSALES

Hyellaceae (thallus organised like filament)

1 Filaments penetrating calcarious substates — *Hyella*
1 Filaments partly or out penetrating substrate — ... 2
2 Thallus made of parallel each filaments — *Scopulonema*
2 Thallus made of radial filaments — ... 3
3 Long erect branches — *Hydrococcus*
3 Short erect branches — *Xenococcus*

Pleurocapsaceae (thallus forming attached flat colonies) *Myxosarcina*

NOSTOCALES

Oscillatoriaceae (trichome made of uniseriate undifferentiated cells)

1 Trichomes straight — ... 2
1 Trichomes spirally coiled — *Spirulina*
2 Trichomes with a prominent sheath — ... 3
2 Trichomes without sheath — *Oscillatoria*

3 Sheath mucilaginous, filamentous forming a thallus with confluent sheath — *Phormidium*
3 Sheath very firm enclosing single or bundle of trichomes ... 4
  4 Trichomes single — *Lyngbya*
  4 Trichomes in bundle — *Microcoleus*

Nostocaceae (trichomes differentiated into vegetative cells heterocysts and spores)

1 Trichomes without firm sheath, generally not endophytic; heterocysts present ... 2
1 Trichome with a firm sheath, single within a sheath; heterocysts present ... 6
  2 Intercalary heterocysts generally in pairs — *Anabaenopsis*
  2 Intercalary heterocysts generally single ... 3
3 Heterocysts commonly terminal with a single large spore adjoining — *Cylindrospermum*
3 Heterocysts terminal rarely; generally intercalary ... 4
  4 Filaments single or in formless gelatinous mass — *Anabaena*
  4 Filaments generally in definite colonies ... 5
5 Thallus finger shaped, attached at first — *Wollea*
5 Thallus otherwise — *Nostoc*
  6 Cells very short, discoid — *Nodularia*
  6 Cells not discoid — *Aulosira*

Scytonemataceae (filaments exhibiting false branching or pseudobranching)

1 Heterocysts absent; apices of trichomes as broad as the rest — *Plectonema*
1 Heterocysts present; single trichome in a sheath ... 2
  2 Apex not tapering — *Scytonematopsis*
  2 Apex not tapering; sheath mostly with parallel lamellation ... 3
3 False branches usually geminate — *Scytonema*
3 False branches usually single and often arising next to a terminal heterocyst — *Tolypothrix*

Rivulariaceae (filaments broader at base and mainly tapering)

1	With terminal heterocysts; filamentous in a spherical or hemispherical thallus	... 2
1	With terminal heterocysts; filamentous free, simple or distinctly false branched dichotomously; corymbose thallus	*Calothrix*
2	Spores not known	*Rivularia*
2	Spores commonly found, single, large	*Gloeotrichia*

STIGONEMATALES (True branched heterotrichous habit) (Nostochopsidaceae, Mastigocladaceae, Mastigocladopsidaceae and Stigonemataceae)

1	With lateral or reverse 'V' shaped branching, pedicellate heterocysts present	... 2
1	With lateral or reverse 'V' shaped branching, pedicellate heterocysts absent	... 3
2	Branching true and lateral	*Nostochopsidaceae*
2	Branching reverse 'V' shaped	*Mastigocladopsidaceae*
3	Branching reverse 'V' shaped, filaments not ending in a hair (Mastigocladaceae)	*Mastigocladus*
3	Branching true and lateral (Stigonemataceae)	... 4
4	Older filaments with many rows of cells; filaments prostrate without a distinct erect system	*Stigonema*
4	Older filaments with a single row of cells or many rows only for short portions	... 5
5	Lateral branches not much different from the main filament	... 6
5	Lateral branches distinct from the main filament	*Fischerella*
6	Hormogones present	*Hapalosiphon*
6	Hormogones not known, endospores present	*Westiellopsis*

Taxonomic studies on Blue-Green Algae

Most of the taxonomic investigations on Blue-green Algae were mainly on floristic surveys of natural populations that included description of taxa and enumeration of available species. The first record of a Blue-green Algae from India was 150 years ago from Assam. *Calothrix indica* was described by Montagne and the first Indian to record blue-green algae is Kirtikar in 1886. Significant contribution on the blue-green algal flora of India were made by Banerji (1935), Rao (1938), Gonzalves & Gangla (1949) and Mitra (1951). The monograph of Cyanophyta by Desikachary (1959) is regarded as one of the classical manuals. Other important contributions on the floristics of blue-green algae are those of Pandey (1965), Tiwari (1972, 1975), Laloroya and Mitra (1973) Anand and Shanthakumar Hopper (1986) and Anand (1990). A comprehensive account of blue-greens of Indian soils was prepared by Subbaraju (1972) and amended recently (see Bongale & Bharathi (1985). Table 1 enumerates the contribution on the survey of blue-greens in rice fields of our country.

Table 1. Literature of taxonomy of blue-green algae of rice fields in India (after Anand, 1990)

States	References
Andhra Pradesh	Laloraya & Mitra, 1973.
Bihar	Laloraya & Mitra, 1973; Jha *et al.*, 1986.
Goa	Bongale, 1981.
Himachal Pradesh	Kamat, 1970.
Jammu & Kashmir	Goyal *et al*, 1984; Khan, 1957.
Karnataka	Tiwari, 1972, 1975; Tiwari & Pandey, 1976; Kamat, 1979; Bongale & Bharathi, 1980; Bongale, 1981, 1984; Shivaprakasa & Shivappa Shetty, 1983; Bongale, 1987.
Kerala	Parukutty, 1940; Aier, 1965; Verma, 1965, Anand & Shanthakumar Hopper, 1987.
Madhya Pradesh	Tiwari, 1972, 1975; Mishra & Purohit, 1979.
Maharashtra	Gonzalves & Gangla, 1949, Marathe, 1963; Marathe & Navalkar, 1963; Sardeshpande & Goyal,1981; Tarar and Shewale, 1984; Kolte & Goyal, 1985; Patil and Satav, 1986.
Meghalaya	Reddy, *et al.*, 1986.
Orissa	Rao, 1938; Laloraya & Mitra, 1973.
Punjab	Sarma & Swarnakanta, 1978; Grover & Pandhol, 1975; Sarma *et al.*, 1979.
Rajasthan	Gupta & Kumar, 1968; Gupta, 1972.
Tamil Nadu	Khurana & Venkataraman, 1968; Tiwari, 1972, 1975; Laloraya & Mitra, 1973; Tiwari & Pandey, 1976; Anand & Revathy, 1987.
Uttar Pradesh	Mitra, 1951; Gupta, 1957; Pandey, 1965; Gupta, 1966; Singh & Chaturvedi, 1970; Prasad & Mehrotra, 1980; Pal & Yadav, 1974; Khan & Mathur, 1976; Prasad & Srivastava, 1968.
West Bengal	Laloraya & Mitra, 1973; Mukhopadhyay & Chatterjee, 1981; Pal & Santra, 1982; Chatterjee & Chatterjee, 1983; Sinha & Mukherjee, 1984; Maity & Santra, 1985; Pal & Santra, 1985.

There have been intensive investigations on individual taxa as well. Bharadwaja (1933) studied the process of false branching in Scytonemataceae and revised the nomenclature of several taxa. He found that the false branching adjacent to single or two pored heterocyst determines the diagnosis of *Tolypothrix* and *Scytonema* respectively. Anand and Gunaseeli (1978), based on their observation under culture condition, questioned the validity of the two genera under nitrogen rich nutritional condition when heterocysts are rarely formed. Occurrence of special type of true branching called reverse V shaped branching was established by Iyengar and Desikachary (1946). This formed the basis for the creation of a family Mastigocladaceae. Other additions to Stigonematales are the establishment of new taxa, *Camptylonemopsis* (Desikachary 1948), *Iyengariella* (Desikachary 1953) and *Westiellopsis* (Janet 1941). Flexibility of the morphological characters used in the delineation of *Stigonema, Fischerella, Hapalosiphon* and *Westiellopsis* were reported by Anand and Sasthankutty Pillai (1980). Jeeji Bai (1977) observed that one isolate of *Calothrix* completely lost its identity in culture and produced stages resembling *Hammatoidea, Plectonema, Homeothrix* and *Lyngbya*. Another isolate looked at different times like *Leptobasis, Tolypothrix* and *Fortiea* besides producing 'Lyngbyacean' filaments suggesting the problems in identifications due to morphological variability. Studying several strains of *Anabaenopsis* for a taxonomic analysis Jeeji Bai *et al.* (1977) consider many as synonyms or morphological forms and an earlier report creating a new genus *Cylindrospermopsis* (Seenayya and Subbaraju 1972) elevating *Anabaenopsis raciborskii* is proved to be unjustified. Anand (1988) reviewed the results of number of studies on morphological variability of blue-green under culture condition and stressed the need for establishing stable characters for correct identifications of taxa. Other results on experimental taxonomy are dealt at length in a later chapter.

I thank the Director, CAS in Botany, University of Madras, for encouragement.

REFERENCES

Aier, R.S. 1965. Comparative algological studies in the rice fields in Kerala State. *Agric. Res. J.Kerala*, **3**: 100-104.

Anand, N. 1988. Culture studies and taxonomy of blue-green algae, certain identification problems. *Arch. Hydrobiol. Suppl.* **80**: 141-147.

Anand, N. 1990. *A hand book of blue-green algae.* Bishen Singh Mahendra Pal Singh, Dehra Dun, pp 1-79.

Anand, N., Gunaseeli, R. 1978. Taxonomic studies on two blue-green algae, *Scytonema* Ag. and *Tolypothrix* Kuetz. *J. Madras Univ.* Pt. B **41**: 138-148.

Anand, N. and G. Revathi. 1987. Blue-green algae from rice fields of Tamil Nadu. *Phykos* **26**: 17-21.

Anand, N. and A. Sasthankutty Pillai 1980. Taxonomic studies on some Stigonemataceae. *J. Madras Univ.* **43**: 18-26.

Anand, N. & R.S. Shanthakumar Hopper 1987. Blue-green algae from rice fields in Kerala State, India. *Hydrobiologia* **144:** 223-232.

Banerji, J.C. 1935. On algae found on soil samples from alluvial paddy fields of Faridpur, Bengal. *Sci. Cult.* **1:** 289-299.

Bongale, U.D. 1981. On the soil algae from paddy fields of Panjim (Goa) and Chikkamanchali (Raichur District, Karnataka), India. *J. Indian bot. Soc.* **60:** 326-329.

Bongale, U.D. 1987. *Phormidium hyalinum* sp. nov. and varieties of *Polychlamydum varium* Ghose from paddy fields of Karnataka. *Phykos,* **26:** 32-35.

Bongale, U.D. and S.G. Bharathi 1980. Algal flora of cultivated soils of Karnataka State, India. *Phykos,* **19:** 95-109.

Bongale, U.D. and S.G. Bharathi 1985. Blue-green algal flora of the Indian soils. *Indian Rev. Life Sci.* **5:** 163-204.

Chatterjee, M. and S.P. Chatterjee 1983. Nitrogen fixing Cyanobacteria from the paddy field soils of Burdwan district. *Phykos* **22:** 64-66.

Desikachary, T.V. 1948. On *Camptylonema indicum* Schmidle and *Camptylonemopsis* gen. nov. *Proc. Indian Acad. Sci.* B, **28:** 35-50.

Desikachary, T.V. 1953. *Iyengariella tirupatiensis* gen. et sp. nov. from South India. *Phytomorphology* **3:** 249-253.

Desikachary, T.V. 1959. *Cyanophyta.* Indian Council of Agric. Res., New Delhi, pp. 686.

Drouet, F. 1951. *Cyanophyta. In: Manual of Phycology,* (Ed.) G.M. Smith, Chronica Botanica, Waltham, Mass, pp. 159-166.

Drouet, F. 1968. Revision of the classification of the Oscillatoriaceae. *Monogr. Acad. nat. Sci. Philad.* **15:** 1-334.

Drouet, F. 1973. *Revision of the Nostocaceae with cylindrical trichomes.* Hafner Press, N.Y.. 1-292.

Drouet, F. 1978. Revision of Nostocaceae with constricted trichomes. *Beihefte zur Nova Hedwigia* Heft **47:** 1-258.

Drouet, F. and W.A. Daily 1956. Revision of the coccoid Myxophyceae. *Butler Univ. Bot. Studies* **12:** 1-218.

Fogg, G.E., W.D.P. Stewart, P.Fay and A.E. Walsby 1973. *The Blue-green algae.* Academic press,.London, England.

Geitler, L. 1932. Cyanophyceae in Rabenhorst's Kryptogamen flora. Leipzing, 1-1196.

Gonzalves, E.A. and K.S. Gangla, 1949. Observations on the algae of paddy field soils. *J. Univ. Bombay* 18: 51-59.

Goyal, S.K., B.M. Sharma and R.S. Gupta 1984. Algal flora of rice field soils of Jammu and Kashmir State. *Phykos,* 23: 59-64.

Grover, I.S. and R.K. Pandhol 1975. Algal flora of paddy fields of Ludhiana and its adjacent areas. *Phykos,* 14: 89-97.

Gupta, A.B. 1957. The algal flora of some paddy fields and its importance in soil economy. Part I. *J. Res.* 4: 1-24.

Gupta, R.S. 1972. Blue-green algal flora of Rajasthan. *Nova Hedwigia,* 23: 481-489.

Gupta, R.S. and H.D. Kumar, 1968. Blue-green algal flora of Udaipur and its neighbourhood. *Revue algol.,* 9: 91-103.

Iyengar, M.O.P. & Desikachary, T.V. 1946. *Mastigocladopsis jogensis* gen. et sp. nov., a new member of the Stigonemataceae. *Proc. Indian Acad. Sci.* 24: 55-59.

Janet, M. 1941. *Westiellopsis prolifica* gen. et sp. nov., a new member of the Stigonemataceae. *Ann. Bot. London* ns 5: 167-170.

Jeeji Bai, N. 1977. Morphological variation of some species of *Calothrix* and *Fortiea.* *Arch. Protistenk* 119: 367-387.

Jeeji Bai, N. Hegewald E. Soeder, C.J. 1977. Revision and taxonomic analysis of the genus *Anabaenopsis. Arch. Hydrobiol. Suppl.* 51: 3-24.

Jha, M.N., U.N. Jha, N. Ahmab and M.K. Mallik. 1986. Cyanobacterial Flora of Pusa and its adjoining areas. *Phykos.* 25: 97-101

Kamat, N.D. 1972. Oscillatoriaceae of Mysore State. *Phykos,* 11: 59-63.

Kamat, N.D. 1974. The Nostocales of the Mysore State. *Phykos* 13: 33-37.

Khan, S.I. 1957. Cultures of the algae of rice-fields of Kashmir. *Biologia* 3: 29-43.

Khan, M. and A. Mathur 1976. Algal flora of the rice fields arounds Dehra Dun, India. *Rev. Algol,* N.S., XI: 333-337.

Khurana, A.S. and G.S. Venkataraman 1968. Algal and fungal flora of paddy soils. *Indian J. Microbiol.,* **8:** 91-94.

Kirtikar, K.R. 1886. A new species of alga, *Conferva thermalis* Birrwoodi. *J. Bombay nat. Hist. Soc.* **1:** 135-138.

Kolte, S.O. and S.K. Goyal 1985. Distributional pattern of Blue-green algae in rice fieldsoils of Vidarbha region of Maharashtra State. *Phykos,* **24:** 156-162.

Laloraya, V.K. and A.K. Mitra 1973. Studies on the blue-green algae of the paddy fields of India. Part I-Cultural studies, general considerations and distribution pattern of the blue-green algae in paddy fields of India. *Nova Hedwigia,* Beiheft, 47.

Maity, H. and S.C. Santra 1985. Blue-green algal flora of 24 Parganas, W. Bengal (India). *Phykos,* **24:** 46-51.

Marathe, K.V. 1963. A study of the effect of fertilizers on the subterranean algal flora of paddy field soils from Karjat. *J. Univ. Bombay,* **31:** 1.

Marathe, K.V. and B.S. Navalkar 1963. Subterranean algal flora of paddy field soils. *J. Univ. Bombay,* **31:** 11.

Mishra, G.P. and Jaishree Purohit 1979. A preliminary list of soil algae from Sagar. *Phykos,* **18:** 63-67.

Mitra, A.K. 1951. The algal flora of certain Indian soils. *Indian J. Agric. Sci.* **21:** 357-373.

Mukhopadhyay, A. and P. Chatterjee 1981. A check list of blue-green algae from the paddy fields of 24 Parganas and Howrah districts of West Bengal-I. *Phykos,* **20:** 81-84.

Pal, T.K. and S.C. Santra 1982. Contribution to the Cyanophyceae of Murshidabad. *Phykos,* **21:** 150-152.

Pal, U.C. and S.C. Santra 1985. Algal flora of Midnapore, West Bengal (India) I. Cyanophyceae..*Phykos,* **24:** 12-17.

Pal, S. and A.K. Yadav 1974. Some Cyanophyceae from Sardhanpur district, a taxonomic enumeration. *Phykos,* **13:** 38-47.

Pandey, D.C. 1965. A study on the algae from paddy field soils of Ballia and Ghazipur districts of Uttar Pradesh, India. part II (A) - Taxonomic considerations, Cyanophyceae. *Nova Hedwigia* **10:** 177-209.

Parukutty, P.R. 1940. The Myxophyceae of the Travancore state, India. *Proc. Indian Acad. Sci.* B **11:** 117-124.

Patil, P.L. and S.D. Satav 1986. A study on Nitrogen fixing Blue-green algae from rice fields of western Maharashtra. *Phykos,* **25:** 113-116.

Prasad, B.N. and R.K. Mehrotra 1980. Blue-green algae of paddy fields of U.P. *Phykos,* **19:** 121-128.

Prasad, B.N. and P.N. Srivastava 1968. Systematic and ecological studies on algae of alkaline (Usar) soils. *Phykos,* **7:** 102-111.

Rao, C.B. 1938. The Myxophyceae of the Madras Presidency, India. *J. Indian bot. Soc.* **17:** 81-96.

Reddy, P.M., Yamnam, D.D. and Imchen, T.Y. 1985. Investigations on the Blue-green algae of North East India. Distribution and Habitat preferences. *Phykos,* **25:** 148-158.

Sardeshpande, J.S. and S.K. Goyal 1981. Distributional pattern of blue-green algae in rice field soils of Konkan region of Maharashtra State. *Phykos,* **20:** 102-106.

Sarma, A., Swarna Kanta 1978. Algal flora of Patiala and its environs, I. *Phykos,* **17:** 105-111.

Sarma, A., Swarna Kanta and Sunita 1979. Algal flora of Patiala and its environs-Cyanophyceae II. *Phykos,* **18:** 13-19.

Seenayya, G. and Subbaraju, N. 1972. On the ecology and systematic position of the alga known as *Anabaenopsis raciborskii* (Wolosz.) Elenk. and a critical evaluation of the forms described under the genus *Anabaenopsis.* In: *Taxonomy and Biology of Blue-green Algae.* Desikachary T.V. (ed), Univ. Madras, pp. 52-57.

Shivaprakasha, M.K. and K. Shivappa Shetty, 1983. Distribution of N_2 fixing Blue-green algae in paddy soils of Shimoga district of Karnataka and their growth rate and N_2- fixing capacity. *J. Soil. Biol. Ecol.,* **3:** 77-82.

Singh, K.P. and U.K. Chaturvedi 1970. Myxophyceae of the Rohilkhand Division, U.P., India. II. *Phykos,* **9:** 36-40.

Sinha, J.P. and D. Mukherjee 1975a. Blue-green algae from the paddy fields of Bankura district of West Bengal-I. *Phykos,* **14:** 117-118.

Tarar, J.L. and T.H. Shewale 1984. Studies on the effects of some fungicides on soil algae of paddy field. *Phykos,* **23:** 191-201.

Tiwari, G.L. 1972. Study of blue-green algae from paddy field soils of India. *Hydrobiologia* 29: 335-350.

Tiwari, G.L. 1975. A study of the blue-green algae from paddy field soils of India II. Taxonomic considerations of non heterocystous BGA. *Nova Hedwigia* 26: 765-797.

Tiwari, G.L. and R.S. Pandey 1976. A study of the BGA from paddy field soils of India III. Nostacaceae. *Nova Hedwigia*, 27: 701-730.

Verma, K. 1965. Culture studies on some members of Chroococcales. *Phykos*, 4: 3-9.

Ind. Phyco. Rev. 1993.
2: 19-30.

BIOSYSTEMATICS

V. SANKARAN
Department of Botany, N.G.M. College : Pollachi - 642 001.
INDIA

The blue-greens constitute a group of organisms exhibiting a very high degree of variability both in nature and culture. The taxonomy of this group was based on preserved and herbarium material. Classical treatises and treatments depended solely on these materials, which formed the basis of such descriptions and reports. Though variability in blue-greens was a known factor, it did not attract the serious attention of blue-green algal taxonomists. There was the least agreement as to the number of orders and families that would accommodate the genera. On one extreme there were splitters (Geitler, Elenkin, Desikachary) and on the other, there were lumpers (Drouet and Daily). The splitters stuck to the classical concepts by maintaining the traditional number of groups and other taxa. The lumpers cut down the number of genera and species drastically on the basis of nomenclatural synonymy. Generic and specific identifications were also dubious due to lack of proper taxonomic circumscription.

Stimulated by culture studies of coccoid blue-greens and also earlier knowledge on variability in blue-greens, and further awakened by the drastic revisions of the coccoid forms by Drouet and Daily, a clarion call was given by Desikachary (1970) for testing the taxonomic concepts by experimental, autecological and synecological, i.e. biosystematic investigations.

Such studies have been undertaken subsequent to his call in our country. Some earlier investigations on similar lines are also known. Indian studies pertaining to the behaviour of the blue-greens both in nature and culture undertaken with a view to elicit maximum information on the taxa concerned, are included in this review on biosystematics. These studies add to the information already available in the classical manuals and other treatments about these taxa.

Attempts at investigating blue-greens in nature in a detailed manner were made by earlier workers. For example, Singh (1942) based on autecological studies, differentiates five morphological expressions in *Wollea bhardwajae:* (1) Common cylindrical sheath attached to the mud, enclosing numerous parallel trichomes. (2) As above, but free floating in water. (3) Contorted and entangled trichomes, enclosed within a ball like mucilage colony. (4) Individual trichomes enclosed in a sheath. (5) Free sheathless trichomes. He considers this genus as a synthetic one and transfers four of the earlier recognised species of this genus to *Anabaena*. But this is a field study spread over a number of years, and not a study under culture conditions in defined media.

Singh (1942) reports four morphological variants in *Raphidiopsis indica* occurring naturally, each variant predominating in one season of the year. These variants are: (1) Straight trichomes with tapering ends. (2) Straight trichomes tapering at one end. (3) Straight trichomes with both ends rounded. (4) Curved trichomes with rounded ends.

Singh (1962) has made observations on the seasonal variants of *Anabaenopsis raciborskii* Wolosz. the alga showed a preference for high organic matter, and low redox potential, growing at pH 7.8-8.5. Heterocyst formation was late. There was a concomitant disappearance of gas vacuoles at this stage. During the course of development the alga was found to pass thro' stages of five different morphological variants that are listed below:

1) Straight trichomes with both ends tapering.
2) Straight trichomes with one end tapering, the other end rounded.
3) Curved trichomes with pointed ends.
4) Trichomes with heterocyst at one end.
5) Trichomes with heterocysts at both ends.

Variants 1 and 3 are similar to *Raphidiopsis indica* Singh. Variants 4 and 5 similar to *Cylindrospermum.* These variants have been shown to be seasonal in distribution.

This investigation gives useful information about habitat parameters and seasonal variations of just one alga only. Such details encompassing a large number of forms will be more fruitful.

Attempts at interpreting information for taxa in an objective manner are also useful. Jeeji Bai (1977, 1980), has employed statistical methods for the analysis of the morphological characters, traditionally used at the specific level in the genus *Anabaenopsis.* She has used statistical average and standard deviation for the length and breadth ratio for the vegetative cells, heterocysts and spores. Measurements from direct observations, and original descriptions have been used. Only those species with constricted and spirally coiled trichomes have been analysed in this exercise. In analysing the planktonic collections, the chemical composition of the lakes has also been considered. Based on averages and standard deviations, she assigns a set of three collections to *A. elenkenii*, two collections to *A. kleifii* and one to *A. intermedia* (all from Hungary). As there is overlapping of characters they are considered more as forms of *A. elenkenii.* Another species from Hungary, *A. arnoldii* is maintained as a distinct one in spite of similar overlapping characters. The collection from Peru is also assigned to a variety of the above mentioned species.

In making taxonomic conclusions out of the investigation, Jeeji Bai points out that *A. arnoldii* var *indica* is identical with *A. magna* which is considered as a "good species".

Another species considered "good" is *A. arnoldii.* She considers the ground for separating the three forms of this species and also other species *A. teodorescii* and *A. millerii* from *A. arnoldii* as not strong. She also considers certain other varieties of *A. arnoldii* indistinguishable from it. *A. elenkinii* is another good species recognised by Jeeji Bai. She suggests that the overlapping characters which make some of the species indistinguishable from others as mentioned above could be resolved if mean values and standard deviations are calculated from a large number of variations, thereby excluding the extreme values. This is purely a recommendatory suggestion. This work differs from conventional taxonomic treatment in its statistical approach and consideration of ecological factors.

Analysis of the cultural behaviour of the blue-greens has been undertaken for different taxa at the University of Madras. Certain strains of coccoid blue-green algae have been investigated under identical culture conditions by Padmaja (See Padmaja & Desikachary 1967). Eight isolates of *Anacystis* and two strains of *Synechococcus* were found to behave in a similar manner under idendical culture conditions, but in different culture media. This study led to the separation of two forms that were morphologically and biochemically distinct from each other. Based on her studies, Padmaja comments that what has been named as *"Anacystis nidulans"*, is perhaps *Synechococcus elongatus*. Earlier studies at the Madras Lab have brought out the involution stages in coccoid forms and heterocystous forms in culture.

Jeeji Bai (1976) has reported on the morphological variations of *Scytonema stuposum* in culture. The alga was isolated in unialgal condition in samples collected from coastal sands of Madras. Five different culture media viz., Chu 10, Zehnder and Gorham media 11 and 32, Allen and Arnon and Gerloff et al. media have been employed for growing the forms. The material grown in Zehnder and Gorham 11 for 15 days was used as the inoculum. All culture were grown in test tubes at room temperature (28^0 C) and illumination 2000 lux for 16-18 hrs. per day. Both liquid and agar cultures were used.

Normal *Scytonema* condition with single as well as geminate false branches with intercalary heterocysts was observed in Chu 10 medium. Involution forms involving broad filaments and crowding of cells were observed in 28 and 35 day old cultures. Typical *Scytonema* conditions were also observed in Gerloff et al. medium. In the Zehnder and Gorham 11 medium, intercalary heterocysts were less common and involution forms were more pronounced.

Scytonema condition tended to become more obscured in the Allen and Arnon and Zehnder and Gorham medium 32. In the Allen and Arnon medium, long unbranched filaments, with occasional short erect irregular projections were observed. Normal branches were completely suppressed. Suppression of heterocysts, cell elongation and involution stages were also observed. Sheath formation and involution forms were much pronounced in Zehnder and Gorham medium 32. Increased cell size, suppression of heterocysts and branches are attributed by the investigator to the increasing concentrations of the constituent inorganic nutrients of the culture medium such as Zehnder and Gorham and Allen and Arnon. The morphology and morphometry

are fairly comparable between isolates grown in culture media with a relatively low concentration of salts and the species descriptions of Geitler. Suppression of heterocysts also disturbs the normal development of the false branches in concentrated media, obscuring the typical morphology of the alga. Employing a wide variety of culture media, thus brings out the range of morphological variation of the alga.

Using similar culture techniques, media and condition as above, Jeeji Bai (1977) has also investigated the morphological variations of *Calothrix* and *Fortiea* species. Two isolates of *Calothrix* have been obtained from the sandy soils of Madras coast and one isolate of *Fortiea* obtained from paddy field soil of a village near Madras. In all the media except the Allen and Arnon medium, the first isolate tentatively identified as *C. membranacea* Schmidle showed typical *Calothrix* condition viz., the trichome with broadened basal part, narrow meristematic apical part, basal and median heterocysts - single or double, and hormogones with tapering ends, the single false branches developing in old cultures. In the Allen and Arnon medium, the typical *Calothrix* condition was lost, with a total loss of heterocysts, developing of *Plectonema* like false branches (in old culture) and loop like twisting of trichomes within the sheath due to intense cell divisions. But attenuated ends were always visible.

The second isolate identified as *C. fullerbornii* Schmidle was found to be more polymorphic. Typical calothricaceous condition was observed in the Chu 10 medium. Plectonemoid false branches were seen (in old cultures) of Zehnder and Gorham 11 medium, twisting of the filaments was observed in the Zehnder and Gorham 32 medium. *Camptylonema* stages with single intercalary heterocysts and tapering ends were seen in Gerloff et al. medium. Extreme changes were observed in the Allen and Arnon medium. These included prolonged elongation and meristematic activity, variously twisted and branched filaments and loss of polarity, attenuation and heterocysts. The filaments were often lyngbyaceous. The isolate assigned to *Fortiea inserta* Skuja showed a distinct broadening of apical portion and intercalary heterocysts - described as *Leptobasis* condition in Chu 10, Zehnder and Gorham 11 and 32, and Gerloff et al., with *Tolypothrix* like stages being more common. In the Allen and Arnon medium, heterocysts and polarity disappeared, resulting in lyngbyaceous condition. The morphological variation of the above three isolates affects the diagnostic characters used at generic, specific and family levels. The tapering and polarity appear to be fairly constant in *Calothrix* isolates, whereas false branching differs in extent.

These investigations in a way cloud the taxonomic identity of the isolate due to a high degree of polymorphism exhibited by them in culture. The traditional taxonomic distinction and circumscription become problematic. The alga in extreme condition simulates another alga with which it is otherwise distantly related.

Jeeji Bai (1985) has made investigations of an experimental taxonomic nature in cultures of *Spirulina fusiformis* Woronich. She has reported the occurrence of two morphones or morphological forms, viz., straight trichome morphone "L" and normal helical trichomes - morphone "S". Study on the morphology and growth preferences of these two morphones in pure and mixed populations have been made in the laboratory, and open air culture conditions. In mixed cultures, population dynamics was found to be due to the competition between the two morphones. These studies pertain to an isolate from Madurai - Tamil Nadu. The cultures have been grown in Zarrouk medium and also as pot cultures (open air), with different chemical treatments.

In initially pure culture of "S" morphones a few "L" morphones were observed after prolonged growth in lab cultures when media transfer was inconsistent. These mixed population showed shifting balance between the two morphones in open air culture. In the open air cultures, a complete exclusion of one or the other morphones never occurred. But in lab cultures, in mixed populations under nutrient deficiency, there was a complete or near complete elimination of "S" morphones, whereas there was a greater proportion of "S" forms under nutrient sufficiency or nutrient excess. In pure populations, with careful monitoring both "S" and "L" forms remained pure under all conditions. Thus the conditions leading to the change in the morphology and behaviour of the two morphones are very interesting.

The change from "S" to "L" is considered to be unidirectional and irreversible and that morphological transformation of S to L is mutation affecting a few trichomes under certain growth conditions. These variants are not known in natural populations. It will be difficult to distinguish the "L" morphone of *Spirulina* from a trichome of *Oscillatoria*. In that case helicity as a taxonomic criterion has to be questioned.

Paromita Bose et al. (1971) report on the ability of certain blue-green algae in culture to solubilize tricalcium phosphate. *Tolypothrix tenuis, Aulosira fertilissima*, strains of *Nostoc* and *Anabaena* which possess nitrogen fixing ability have been found to solubilize tricalcium phosphate. They, thus possess ability to dissolve bound phosphate. Besides they have been found to produce growth substances and also to tolerate high levels of agro-chemicals. Other cultures and species of the above taxa may possibly exhibit varying levels of solubilization and tolerance. Such findings will be useful not only from the point of view of taxonomy, but also from that of practical application.

Workers started going in for non-conventional characters. Consequently cultures were studied for their tolerance for various chemical and a wide variety of other compounds and conditions.

Singh (1973) has studied the tolerance of certain blue-green algae towards selected pesticides. The nitrogen fixing *Cylindrospermum* sp., *Aulosira fertilissima* and non fixing *Plectonema boryanum* were tested for their ability to tolerate benzene hexachloride, lindane, diazinon and endrinapon. The cultures were grown in Chu 10 medium and the concentration of the pesticides used was upto 500 ug/ml. The studies revealed various degrees of tolerance of the cultures to the different pesticides used. The cultures were found to be more sensitive to benzene hexachloride than to other pesticides used in the investigation. *Aulosira fertilissima* and *Plectonema boryanum* were found to be more resistant to the pesticides than other cultures.

Similar studies have been carried out by Ahmed and Venkataraman (1973). They have studied the tolerance of *Aulosira fertilissima* towards pesticides including MCPA, MEPB, Stain F-34, Lindane, parathion, endrin, diazinone, BHC, sevin, which are used at 1-2 ppm in paddy cultivation. The alga exhibited high levels of tolerance towards these pesticides.

It has become necessary, in course of time to study a very large number of forms and to analyse both the conventional as well as the non-conventional characters, taking into consideration as many characters as possible, and giving equal weightage to all of them. These studies were based on Adansonian principles. Such large numbers of characters and taxa needed the use of computer for their handling. The beginnings were made in the Department of Botany, University of Madras.

Sankaran (1972) has studied several cultures of *Nostoc* obtained from different culture collection centres of the world. These cultures were studied under identical conditions. Initially 39 cultures of *Nostoc* were analysed for their morphological variations in three culture media, viz., modified Chu 10 medium, Allen and Arnon medium, and the biphasic soil water medium. Details of plant mass nature, length, width and length/width ratio of the vegetative cells, heterocysts and spores, inter heterocystous distance, spore : vegetative cell ratio were the morphological characters studied. These were coded as per standard procedures. Treating each isolate as an operational taxonomic unit - OTU, the similarity between them was calculated by correlation coefficients and taxonomic distance. (See Sokal and Sneath 1963). This procedure was carried out with the aid of CDC3600 computer. For the same isolates more number of characters was elicited by subjecting them to various treatments under identical culture conditions.

The characters are given below :

1. Level of combined nitrogen source needed for heterocyst suppression.
2. Spore forming ability in the presence of nitrogen, phosphorus and sulphur compounds.
3. Increased cell size (length and width) in response to addition of varying amounts of nitrate, sulphate, chloride and phosphate.
4. Ability for growth in N-free medium.

5. Tolerance towards high levels of nitrogenous compounds and sulphates.
6. Tolerance towards pH, salinity, urea, thiourea and copper sulphate.

These culture experimental results yielded 34 more non-conventional characters in addition to the morphological characters already listed. This gave a total of 49 characters for each of the 39 OTU's. A similar computer analysis was carried out for these 39 isolates with respect to the 49 characters. This second analysis brought greater differences between the isolates in terms of similarity estimation than the first analysis. The OTU's which showed greater similarity based on morphological characters, showed greater dissimilarity when they were objectively compared, using additional characters listed above. An objective interpretation was not feasible as no comparison was possible with the species derived from classical taxonomy. Hence subsequent analyses have been made with reference to morphological characters, using such analysis for 123 isolates was the next step.

The 123 isolates of *Nostoc* included a number of isolates traceable to a common source of isolation. In the analysis none of these behaved as identical pairs. They have always maintained a distance from each other. It is to be however mentioned here that some of the isolates in the 39 OTU analysis stood apart, became closer in the 123 OTU analysis. This due was to the increase in the number of the OTU's. In the 123 OTU analysis many isolates have clustered at distance 0. i.e., close similarity level. All these are OTU's from different culture collections and have been identified variously by different workers. This numerical analysis has brought together different isolates, but has separated those that have been considered to be identical. For example two isolates each of *N. piscinale, n. verrucosum* and *N. ellipsosporum,* as identified by Kantz and Bold have stayed apart and have never come close in this analysis.

Yet another type of numerical analysis involved the inclusion of the species of *Nostoc* described in the classical manuals along with the 123 isolates. The characters given for each species in the manuals were coded as per standard procedures, and their taxonomic resemblance estimated. The various OTU's did not get integrated with the species from the manuals. These two tended to remain in different groups. The characters of the culture isolates were in no way comparable with the features used for descriptions of the manuals. The isolates could not fit with the text book species. The clustering analysis did not yield any clear cut groups identifiable with traditional sub-divisions of the genus *Nostoc*. It was not possible to compare the culture with "Species" based on characters derived from behaviour in the culture media. Numerical procedures did not produce any meaningful and interpretable results. The cultures and species remained immiscible and apart in the dendrogram. This behaviour confirmed that species can be considered as single phenotypic expressions and hence did not easily match with composite or cumulative data, obtained for cultures based on expressions under different conditions or multiplicity of expressions.

Prasad (1976) made a comparative study of a larger number of isolates of *Nostoc*. The results of numerical analysis of the text book species of the genus and isolates of the same are presented by Sankaran and Prasad (1980). They have followed a different procedure in this analysis. First 60 taxa of *Nostoc* (55 species and 5 varieties) - TBS (Text Book Species) were analysed by using weighted pair group method (WPGM) on the basis of taxonomic distance (djk) (Sokal and Sneath 1963). The data obtained for 278 isolates were added to the TBS. In the second phase the data obtained for 278 isolates from Chu 10 medium and soil water medium were separately analysed. In both these analyses many OTU's formed discrete clusters of their own, far removed from any TBS. This may be regarded as failure to discover the appropriate conditions in which the isolate may express phenotypic characters similar to those of one or the other species.

It is questionable whether this separation between the culture isolates and TBS is significant enough for the creation of new taxa.

The methodology was modified here in view of the large number of isolates and limited computer facilities at the time of this analysis. The cultures were processed in two sets along with the TBS. All the OTU's in any set that joined at djk=0.0, but one in any clustering were eliminated and the rest were clustered again until each became smaller by recycling. Then they were run as a single set, repeated by similar recycling until no OTU joined at djk=0.0.

The results of cluster analysis made with the addition of characters derived from Chu and soil water medium were not directly translatable into taxonomic conclusions. However, in the case of some species, significant points have arisen which may be used in aiding taxonomic conclusions. Clusters were usually smaller in soil water analysis than in Chu.

Nostoc spongiaeforme, N. paradoxum, N. ellipsosporum, N. commune, N. sphaericum, N. macrosporum, N. verrucosum, N. parmeleoides, N. insulare, N. minutissimum, N. ellipsoideum and *N. linckia* var *arvense* did not unite with either a culture OTU or with another taxon in any of the cluster exercises.

The genus *Nostoc* has received taxonomic treatment by way of grouping of the species at the hands of Bornet and Flahault (1888), Elenkin (1938), Starmach (1966), and Kondrateva (1968). Desikachary (1959) accepts the treatment of Geitler (1932). The present analysis of the 60 commonly known species - TBS, did not support totally any of these classical treatments of the genus. But out of the analysis 5 major groupings have been recognised, which somewhat correspond to the groups recognised by Bornet and Flahault. These are : (1) Paludosum group (13 species), (2) Commune group (25 sp.), (3) Muscorum group (14 sp.), (4) Spongiaeforme group (2 sp.) and (5) Ellipsosporum group (only *N. ellipsosporum*).

When the TBS dendrogram was compared with that involving the culture isolates, clustering pattern was found to be different. Cultures of identical pairs of isolates (Sankaran unpublished) and even clones (Prasad unpublished) did not join at djk = 0.0.

It was also noticed that large number of species did not associate themselves with any culture isolate. Only a few isolates were interspersed among the species.

In the light of the above analysis, it is suggested that more number of isolates of *Nostoc* from different geographical and ecological conditions should be included to have whole array or spectrum of forms intergrading between species. It may then be possible to start experimental taxonomy with known or acceptable culture types to make generic revision possible. Study of hundreds of cultures in different culture media for nearly five years by two investigators still does not permit generic revision and only suggests to keep classical species and cultures aloof. For culture investigations culture descriptions from standard medium such as chu 10 medium is advised so that accumulated data can be processed in future by maintaining environmental conditions and age of culture on par with the present investigation.

Anand (1979) has utilised the experimental data from the 71 isolates of *Anabaena* in yet another way by assigning them to 19 book species and proposing a diagnostic key for these species. Position of spores, whether adjacent to or away from the heterocysts was found to be the only stable character in the species studied. Hence, this character and other spore characters are used in the preparation of the diagnostic key. These characters are from the Chu and the soil water cultures under standard conditions.

The author's identification of certain of the isolates is different from the previous identifications made by earlier workers. For instance, the isolate identified as *A. oscillarioides* by Forest has been identified as *A. ambigua*. Similarly the isolate identified as *A. viguieri* by Kantz and Bold has been reidentified as *A. variabilis*. These are example only out of many instances of reidentifications. Earlier culture studies by Sankaran (1972) and Prasad (1976) on *Nostoc* as stated elsewhere already in this review brought out a similar experience. The diagnostic key provided by Anand is for 19 species. But it is not based on fixed material as in classical taxonomy, but based on culture studies of living material.

Anand (1980) has studied 71 isolates of *Anabaena* in the same pattern followed for *Nostoc* by Sankaran (1972) and Prasad (1976). All the 71 isolates were grown in the Chu 10 medium and the biphasic soil water medium from a standardised inoculum. From this study 37 characters for each isolate were gathered as also for 97 book species based on the descriptions from standard manuals. This made a total of 168 operational taxonomic units (OTU's). Cluster analysis was carried out by using weighted pair group method (WPGM) of the taxonomic distance, (Sokal and Sneath 1963). The results of the analysis revealed that 18 text book species did not form cluster even at the average level, thereby retaining their individuality. This included the following species :

A. oryzae, A. sphaerica, A. batophora, A. laxa, A. circinalis, A. flos-aquae, A. utermohlii, A. echinospora, A. lavenderii, A. delicatula, A. randhawae, A. knipowitschii, A. limnetica, A. affinis, A. salina, A. tatarica, A. hollerbachii and *A. lohamari.*

14 species formed clusters, in which the OTU's were distantly united. 65 species of the text book formed clusters with Chu 10 cultures, and 9 with the soil water cultures. Certain of the isolates were identical in the Chu 10 medium and some in the soil water medium.

Gabriel (1978) carried out numerical analysis of 79 isolates of *Calothrix*, 23 isolates of *Dichothrix*, by using WPGM of djk. His results suggest that the genus *Dichothrix* may not be very distinct from *Calothrix* as was suggested by Elenkin and other workers.

As is well known the blue-greens constitute an extemely large group of organisms that are extremely variable. Studying them in nature and culture will represent two different methodologies. The classical taxonomist describes a form in the traditional way and as suh has his own limitations. The experimental taxonomist isolates them and studies them in culture, working out the variations in structure and behaviour. He depends upon the classical taxonomist for the most appropriate identification. Identifying an alga based on culture characters elicited by the experimental taxonomist is not of any use in proper identification. The classical taxonomist does not go in for more characters. A close and complementary interaction between these two categories of taxonomists is highly essential to work out the degree of morphological plasticity and the very wide range of non-conventional character of a traditionally established taxon and subsequently of its group by experimental procedures and their handling and interpretations by màthematical or statistical means.

REFERENCES

Ahamad, M.H., & G.S Venkataraman, 1973. Tolerance of *Aulosira fertilissima* to pesticides. *Curr. Sci.*, 42: 108.

Anand, N. 1979. Studies on the genus *Anabaena* Bory: key to the identification of some *Anabaena* sp. in culture. *Acta Botanica Indica*, 7: 22-28

Anand, N. 1980. Preliminary studies on a revision of the genus *Anabaena* Bory. *Taxonomy of Algae*. T.V. Desikachary ed. Madras: 109-113.

Bornet, E., & Flahault, C., 1886-1888. Revision des Nostocaces Heterocystees. *Annls. Sci. Nat. Bot. Ser.* VIII. 3: 322-388; 4: 343-375; 5: 51-129; 7: 177-262.

Desikachary, T.V. 1959. *Cyanophyta*. Indian Council of Agricultural Research, New Delhi. pp. 686.

Desikachary 1970. Taxonomy of blue-green algae-problems and prospects. *Schw. Zeitsch. fur Hydrologie*. 32., fasc. 2: 490-94.

Elenkin, A.A. 1938. *Monographia alagarum Cyanophycearumk aquidulcium et tercestrium iηfinibus URSS inventarum pars specialis.* I. Russian Akad. Nauk. SSSR, Moscow and Leningrad, USSR.

Gabriel, M.S. 1978. Studies on the members of *Rivulariaceae* and *Scytonemataceae. Ph.D. thesis.* Madras University.

Geitler, L. 1932. Cyanophyceae in Rabenhorst's kryotogamenflora. *Leipzig* **14: 1196** pp.

Jeeji Bai, N. 1976. Morphological variation of certain blue-green algae in culture: *Scytonema stuposum* (Kutz) Born. *Scheiz. Z. Hydrol.,* **38:** 56-62.

Jeeji Bai, N. 1977. Morphological variation of some species of *Calothrix and Fortiea. Arch. Protistenk. Bd.* **119:** 367-387.

Jeeji Bai, N. 1980. Taxonomic studies on the genus *Anabaenopsis. Taxonomy of Algae,* T.V. Desikachary ed., University of Madras. : 115-135.

Jeeji Bai, N. 1985. Competitive exclusion or morphological transformation. A case study with *Spirulina fusiformis. Arch. Hydrol. Suppl. 71.* **1/2:** 191-199.

Hegeward, E. & J. Soeder. 1977. Revision and Taxonomic analysis of the genus *Anabaenopsis. Arch. Hydrobiol. Suppl.* **51:** 3-24.

Kondrateva, N.V. 1968. *Cyanophyta. Hormogoniophyceae.* Kieve. 522 pp.

Padmaja, T.D. & T.V. Desikachary 1967. Trends in the Taxonomy of the algae. *Bull. National Inst. Sci. India.* No. **34:** 338-364.

Paromitra Bose, V.S. Nagpal, G.S. Venkataraman & S.K. Goyal. 1971. Solubilization of tricalcium phosphate by Blue-green algae. *Curr. Sci.* **40:** 165-167.

Prasad, A.K.S.K. 1976. The Taxonomy of the genus *Nostoc* Vaucher. *Ph. D. Thesis,* Univ. Madras.

Sankaran, V. 1972. Taxonomic studies in the genus *Nostoc* Vaucher. *Ph.D. Thesis,* Univ. Madras.

Sankaran & A.K.S.K. Prasad 1980. Preliminary conclusions on a revision of the genus *Nostoc* Vaucher. *Taxonomy of algae,* T.V. Desikachary ed. University of Madras: 97-113.

Singh, P.K. 1973. Effect of pesticides on blue-green algae. *Arch. Mikrobiol.* 89: 317-320.

Singh, R.N. 1942. A new species of the genus *Raphidiopsis* Fritsch & Rich. (*Raphidiopsis indica* sp. nov.) exhibiting morphological variants. *J. Indian Bot.Soc.* 21: 239-246.

Singh, R.N. 1942. *Wollea Bharadwajae* sp. nov. and its autecology. *Ann. Bot. Lond.* n.s. 6: 593-606.

Singh, R.N. 1962. Seasonal variants of *Anabaenopsis raciborskii* Wolosz. *Hydrobiologia.,* 20: 87-91.

Sokal, R.R. & P.H.A. Sneath., 1963. *Principles of Numerical Taxonomy.* W.H. Freeman & Company, Sanfrancisco & London. pp. 359.

Starmach, K. 1966. . Cyanophyta - Sinice, Glaucophyta. Glaukofity (Polish), *Flora Slodkowodna, Pol. Acad. Sci.* 2: 1-807.

Ind. Phyco. Rev. 1993.
2: 31-40.

REPRODUCTION, DIFFERENTIATION AND DEVELOPMENTAL CYCLES

N. ANAND & G. REVATHY,
Centre of Advanced Study in Botany, University of Madras,
Madras - 600 025.

INTRODUCTION

Blue-green algae, which are truly cosmopolitan in distribution, uniquely lack sexual reproduction, and cell division is the simplest and general mode of multiplication. The sexual phase being totally absent or unknown, propagation is effected also by differentiated vegetative cells and fragments of filaments or colonies. The nature of reproductive structures formed the basis of the classification of blue-green algae. Thuret (1875) divided the blue-green into two tribes, i) Coccogoneae, characterized by unicellular reproductive bodies and ii) Hormogoneae, reproducing by few celled short pieces of trichomes called hormogones. The coccoid members produce different types of reproductive bodies while the filaments members have differentiated cell types, viz., heterocysts, akinetes and the few-celled hormogones.

REPRODUCTIVE STRUCTURES

Endospores, exospores and baeocytes are the reproductive structures produced by coccoid forms. Endospores are formed by successive division of protoplast within a cell and resemble aplanospores of green algae. These are naked spores (sometimes with a wall), which secrete a thin wall when released. Such spores are characteristic of Pleurocapsales. Exospores of all attached forms are serially abstricted by transverse division and are surrounded by a delicate membrane. Sometimes the exospores germinate without being liberated. Baeocytes are produced in Pleurocapsales.

All filaments blue-greens produce hormogones. Pseudohormogonia or hormocysts are seen in members belonging to Scytonemataceae and Stigonemataceae (Desikachary 1959). Hormocysts resemble hormogonia but have a thick pigmented sheath. Hormogonia are motile, but hormocysts are not. Certain filamentous forms have specialized cells called 'heterocysts' which are also implicated with the function of reproduction. Heterocysts may be terminal, basal or intercalary in trichome. Akinetes are large and thick walled resting spores that have accumulated large amount of food reserves in the form of cyanophycin granules or structured granules. Akinetes form the basis of distinction of the species of *Anabaena*, differing in their shape and position in a trichome with reference to the heterocysts (Anand 1979).

Germination of Heterocysts and Akinetes:

Germination of heterocysts is shown to be mode of reproduction in several blue-greens (see, Desikachary 1959). Heterocyst germination has been recorded in *Anabaena, Nostoc, Calothrix, Rivularia, Gloeotrichia, Tolypothrix* and *Brachytrichia* by Kale and Talpasayi (1969). Singh RN and Tiwari (1970) report that germination of heterocyst is a common feature in *Gloeotrichia ghosei. In situ,* germination of heterocysts is also reported by Singh RN *et al.* (1972). Singh HN and Srivastava (1968) reported that spore germination, heterocysts production, hormogone formation and sporulation are the morphological stages of *Anabaena doliolum,* under culture conditions. Heterocyst formation and sporulation were inhibited in the presence of nitrogen source in the medium. Reddy and Talpasayi (1974) studied the heterocyst development in *Cylindrospermum* sp. in the ammonium source and concluded that the polarity of the trichomes are predetermined. Transfer of heterocyst-free trichomes to nitrogen-free medium showed the formation of terminal heterocysts first on one side and later the other end. Anand (1978) reports that heterocysts and spores were inhibited by exogenous nitrogen only in certain *Anabaena* species where the spores are contiguous with heterocysts. Sporulation without heterocyst differentation is seen in *A. variabilis* in which spores are away from the heterocyst. Anand (1980) observed that in culture heterocysts may germinate even when they are attached to the trichomes and also when detached and scattered.

Germination of akinetes is commonly seen when old cultures are transferred to fresh media. Germination of akinetes precedes complete utilization of cyanophycin and cell division occurs before the emergence of the germling. The specific factor responsible for initiation of akinete germination is yet to be unequivocally proved though several environmental conditions under laboratory have been tested (Nichols and Adams 1982). Kaushik and Kumar (1970) concluded that in *Anabaena doliolum* and *Fischerella muscicola* non-photosynthetic wave length of light favoured akinete germination. Biochemical studies on akinete germination by Singh HN and Sunita (1974) in *A. doliolum* by transferring akinetes from dark to light condition revealed the order of macromolecular synthesis during germination of akinetes.

DIFFERENTIATION

The blue-green algal cell has a multilayered cell wall which is about 35-40 mm thick, predominantly made of peptidoglycan and lipolysaccharides. Recently, Adhikari (1985) compared the outermost surface structures of *Chroococcus minutus* and *Gloeothece* sp. In *C. minutus* the sheath consisted of a fibrous material adherent to outer membrane, while in *Gloeothece* the fibrous sheath appeared as two distinct layers. The outer sheath was electron dense consisting of fibrous material running parallel to the cell wall and the inner layer contains a loose network of polysaccharide. The most extensive membrane structure is the thylakoid (flattened vesicles), that appear tripartite, about 14 mm thick. Phycobiliproteins, the principal light harvesting antennae, are located on the thylakoids in granules called phycobilisomes. These pigments constitute nearly 40% of the total soluble protein of the cell. Other inclusions of the cell are polyglucon granules, polyphosphate bodies, polyhedral bodies (200-300 nm diam), cyanophycin granules (500 nm diam) and ribosomes.

Ultrastructural changes during differentiation:

Heterocysts and akinetes (spores), are differentiated vegetative cells and the ultrastructural changes during the differentiation has been understood in several genera. Literature available on ultrastructure of different species is presented in Table 1.

Table 1. Blue-green algal species studied for Ultrastructure of cells, heterocysts and akinetes.

Name	Reference
Anacystis nidulans	Allen 1968, Gannt and Conti 1969.
Synechococcus sp.	Lawry and Jensen 1979
S. lucidus	Edwards *et al.* 1968
Oscillatoria chalybea	Lamont 1969
O. rubescens	Jost 1965
Phormidium unicinnatum	Frank *et al.* 1962
Plectonema boryanum	Jensen 1969, Lawry and Simon 1982
Anabaena azollae	Grilli 1964, Lang 1965
A. circinalis	Fay *et al.* 1984
A. cylindrica	Fay 1969, Walsby and Nichols 1969, Wolk and Simon 1969, Dunn and Wolk 1970, Bryce *et al.* 1972, Roussand-Jacquemin 1983.
A. solitaria	Cmiech and Leedale 1986
A. variabilis	Braune 1980, Simon 1980
Chlorogloea fritschii	Peat and Whitton 1967, Findley *et al.* 1970.
Cylindrospermum sp.	Miller and Lang 1968, Jensen and Clark 1969.
Nostoc sp.	Sutherland *et al.* 1979, 1985.
N. muscorum	Wilden and Mercer 1963, Chao and Bowen 1971.
N. pruniforme	Jensen and Bowen 1961.
Gloeotrichia echinulata	Cmiech *et al.* 1984.

The first indication of a vegetative cell transforming into a heterocyst is the increase in size, followed by the formation of thick envelope. The envelope consists of an inner laminated, outer fibrous and central homogenous layers. The cell wall is predominantly made of carbohydrates that is different from vegetative cells. A reticulate system of evenly distributed membranes replaces the concentric peripheral thylakoids, devoid of biliproteins. Polyunsaturated lipids disappear and glycolipids and polyhydroxy alcohols are found accumulating. Polyphosphate bodies, polyhedral bodies (carboxysomes) and cyanophycin granules disappear. Ribosomes get reduced in number during differentiation of heterocysts. Akinete formation also precedes cell enlargement and thickening of the electron opaque layer. Cell wall of mature akinete is about 80-90 nm thick. Due to heavy deposition on the outer wall, an electron transparent intine and electron dense exine can be distinguished. The akinete envelope comprises of carbohydrates (41%), aminoacid compounds (24%) and lipids (11%) of the total dry weight. In *Anabaena variabilis* thickening of the electron opaque layer is absent. Photosynthetic lamellae get distributed throughout the protoplast and get contorted. Due to extensive elaboration of lamellae the distinction of nucleoplasmic region is completely lost. There is a marked change in the pigment content with chlorophyll being replaced by phaeophytin, increased xanthophyll and absence of phycocyanin, besides reduction of fatty acids up to 85%. Polyglucan and polyhedral bodies and ribosomes become restricted but there is nearly 8 fold increase in the cyanophycin content. Increase in cyanophycin granules is reported to be directly proportional to akinete formation (Sutherland *et al.* 1979).

Physiological and biochemical changes during the differentiation:

Changes in several metabolic processes have been studied during the differentiation of vegetative cells into heterocysts and akinetes (spores) in filamentous blue-greens. Heterocyst formation is generally attributed to depleted nitrogen in the ambient milieu and akinete formation to termination of complete development of *Cylindrospermum* (Anand and Rengasamy 1982). In heterocystous forms appearance of nitrogenase activity coincides with the formation of heterocysts. Further, disappearance of both heterocysts and nitrogenase activity when the nitrogen fixing cells are transferred to nitrogen available condition has been experimentally proved. Protein profiles of isolated heterocysts contain unique components attributed to the enzyme nitrogenase. Heterocysts do not have the oxygen evolving Photosystem II but retain the photosystem I and cyclic phosphorylation. (Stewart *et al.* 1982). Mature heterocysts are devoid of the enzyme Phospho-ribulokinase that generates the substrate for RUBP - carboxylase and several other enzymes involved in carbon metabolism namely, phosphoglycollate, glycollate dehydrogenase (Codd *et al.* 1980) and fructose 1-6, diphosphate aldolase (Lex and Carr 1974). However, in heterocysts the enzymes of oxidative pentose pathway viz., glucose 6-phosphate dehydrogenase and 6-phosphogluconate dehydrogenase are more active. Akinetes lack all systems of nitrogen metabolism and remain as store house of cyanophycin. Singh and Sunita (1974) report a sequence of macromolecular synthesis in germinating akinetes. Synthesis of RNA started at four hours and continued till 20 h, chlorophyll at 8 h, phycocyanin at 12 h and DNA at 24 h.

DEVELOPMENTAL CYCLES

Besides the study of differentiation of heterocysts and akinetes there are certain reports on the development of the thallus under culture condition. A simple life cycle was proposed for *Chlorogloea fritschii* (Fay *et al.* 1964), in which the endospores developed into aseriate condition. Based on this study certain filamentous forms have been investigated.

Lazaroff and Vishniac (1961) describe a photocontrolled developmental cycle of *Nostoc muscorum*. They report a sequential pattern of development of a hormogone first to an aseriate condition and then back to the free trichome condition, in heterotrophic cultures in dark. The free hormogones first differentiate terminal heterocysts while the intercalary cells elongate in a plane parallel to the axis of the trichome before division resulting in a cluster of cells surrounded by a thick sheath (aseriate packets). When the aseriate packets in dark grown cells is exposed to 1000 lux continuous illumination, hormogonia are released and long filaments are seen. The long filaments swarm and appear as spiral aggregates. Lazaroff (1966) concludes that in *N. muscorum* the life cycle is controlled by a red-light triggered, non-photosynthetic induction of differentiation.

Singh RN and Viswanathan (1972) report that in *Camptylonema lahorense* Ghose three major phases of life cycle can be distinguished in culture. These are : 1. Hormogone stage, 2. Juvenile stage, 3. Mature stage. The hormogonia are usually liberated after 18-20 days and remain suspended freely in the medium. A single median heterocyst is differentiated after 42 h; sometimes heterocysts appear in pairs. Juvenile stage is marked by the formation of the crescent shaped filaments and elongation of hormogonia without any branching for nearly three days. Inhibition of branching and formation of colony mark the mature stage that remains for a week. Transfer of the alga to fresh medium results in hormogone release and the cycle follows.

Ladha and Kumar (1975) report several developmental stages in *Oscillatoria* under culture conditions, that were broadly similar to that reported in *Nostoc muscorum.* In cultures raised from single filaments profound changes were observed including branching similar to members of Scytonemataceae and Rivulariaceae and a sequence of developmental stages is as follows: two normal filaments twining around each other forming a rope like aggregate, terminal aggregation of two filaments, formation of packet-like structures due to localized growth in a filament, progress in twining and roping with the aggregate enclosed in a common sheath; apparent fusion of terminal cells within the sheath and dissolution of mucilage envelope resulting in hormogone-release. Kumar and Gupta (1979) report that in *N. muscorum* addition of cyclic nucleotides like cAMP (adenosine monophosphate) to the growth medium effected the aseriate condition of the trichome in light. Aseriate condition was normally obtained in sucrose grown cultures, hence a role for cAMP in aseriate condition is implied.

Development of trichomes in *Cylindrospermum* sp. starting from the germination of the akinete (spores) was studied by Anand and Rengasamy (1982). The trichome developed from the germling, differentiated a terminal heterocyst first on one end and later the other. Spore formation was delayed and similar to heterocyst differentiation. A subterminal spore was found also on one end only. Sometimes instead of a spore the subterminal heterocyst itself is seen. The stage when a trichome possesses two subterminal spores adjacent to the terminal heterocyst with an intervening 60-70 vegetative cells, the trichome appears to be completely developed. *Cylindrospermum* sp. generally do not produce intercalary heterocysts. Once the trichome is fully developed detachment of spores is slowly initiated. Detachment of the first spores was always followed with differentiation of a heterocyst by the immediate next cell, and then a subterminal spore. This process becomes continuous and at one stage there are spores with their terminal heterocysts still attached. Study of the function of terminal heterocysts of *Cylindrospermum* sp. seems essential.

Control and regulation of sporulation in blue-greens is an aspect that has to be understood better. Knowledge on inducing akinete formation (or sporulation) will help to develop efficient biofertilizer programmes, especially if one thinks of distribution of spore dust for dissemination in the field. Ultrastructure of Indian blue-greens is virtually untouched. Ultrastructure of environmentally adapted cells may reveal mechanisms of adaptation.

We thank the Director, CAS in Botany, University of Madras, for all encouragement and Miss R. Gowri for help in Bibliography.

REFERENCES

Adhikary, S.P. 1985. Outermost surface structures in the cyanobacteria *Chroococcus* and *Gloeothece*. *Phykos*, 24-102-116.

Allen, M.M. 1968. Photosynthetic membrance system in *Anacystis nidulans*. *J. Bacteriol*. 96: 836-841.

Anand, N. 1978. Role of heterocysts in sporulation of blue-green algae. *J. Madras Univ.*B 41: 15-19.

Anand, N. 1979. Studies on the genus *Anabaena* Bory II. A key to the identification of some *Anabaena* species in culture. *Acta Botanica Indica*, 7: 22-28.

Anand, N. 1980. Heterocysts as reproductive structures in blue-green algae. *Curr. Sci.*49: 84.

Anand, N. Rengasamy, R. 1982. Differentiation of heterocysts and spores in *Cylindrospermum*. *Curr. Sci.* 51: 942-944.

Braune, W. 1980. Structural aspects of akinete germination in the cyanobacterium *Anabaena variabilis*. *Arch. Mircobiol.* **126:** 257-261.

Bryce, T.A., Welti,D., Walsby,A.E. and Nichols,B.W. 1972. Monohexide derivatives of long-chain polyhydroxy alcohols; a novel class of glycolipid specific to heterocystous algae. *Phytochemistry,* **11:** 295-302.

Chao, L. and Bowen,C.C. 1971. Purification and properties of glycogen isolated from blue-green alga, *Nostoc muscorum*. *J. Bacteriol.* **105:** 331-338.

Cmiech, H.A., Leedale,G.F. and Reynolds, C.S. 1984. Morphological and ultrastructural varibility of planktonic Cyanophyceae in relation to seasonal periodicity. I. *Gloeotrichia echinulata:* vegetative cells, polarity, heterocysts, akinetes. *Br. Phycol. J.* **19:** 259-275.

Cmiech, H.A., and Leedale,G.F. 1986. Morphological and ultrastructural variability of planktonic Cyanophyceae in relation to seasonal periodicity. II. *Anabaena solitaria:* vegetative cells, heterocysts, akinetes. *Br. phycol. J.* **21:** 81-92.

Codd, G.A. Okabe,K. and Stewart,W.D.P. 1980. Cellular compartmentation of photosynthetic and photorespiratory enzymes in the heterocystous cyanobacterium *Anabaena cylindrica*. *Arch. Microbiol.* **124:** 149-154.

Desikachary, T.V. 1959. *Cyanophyta*. ICAR, New Delhi, pp 686.

Dunn, J.H., and Wold,C.P. 1970. Composition of the cellular envelops of *Anabaena cylindrica*. *J. Bacteriol.* **103:** 153-158.

Edwards, M.R., Berns, D.S., Ghiors,W.C., and Holt,S.C. 1968. Ultrastructure of the thermophilic blue-green alga, *Synechococcus lividus* Copeland. *J.Phycol.* **4:** 283-289.

Fay, P. 1969. Cell differentiation and pigment composition in *Anabaena cylindrica*. *Arch. Mikrobiol.* **67:** 62-70.

Fay, P., Kumar, H.D. and Fogg, G.E. 1964. Cellular factors affecting nitrogen fixation in the blue-green alga, *Chlorogloea fritschii*. *J.gen Mircobiol.* **42:** 310-321.

Fay, P., Lynn, J.A. and Majer,S.C. 1984. Akinete development in the planktonic alga *Anabaena circinalis*. *Br. phycol. J.* **19:** 163-173.

Findley, D.L., Walne, P.L. and Holton,R.W. 1970. The effects of light intensity on the ultrastructure of *Chlorogloea fritschii* Mitra grown at high temperature. *J. Phycol.* **6:** 182-188.

Frank, H.,Lefort,M., and Martin, H.H. 1962. Elektronoptische und chemische untersuchungen an Zellwaende der Blaulage *Phormidium uncinnatum. Z. Naturf.* **17:** 262-268.

Fogg,G.E., Stewart,W.D.P., Fay,P., and Walsby,A.E., 1972. *The Blue-green Algae.* Academic Press, London.

Gantt, E. and Conti, S.F. 1969. Ultrastructure of blue-green algae. *J. Bacteriol.,* **97:** 1486-1493.

Grilli, M. 1964. Infrastructure di *Anabaena azollaea* vivente nelle foglioline di *Azolla carolniana. Annali Microbiol. Enzimol.* **14:** 69-90.

Jensen, T.E. 1969. Fine structure of developing polysulphatebodies in a blue-green alga, *Plectonema boryanum. Arch. Mikrobiol.* **67:** 328-338.

Jensen, T.E., and Bowen,C.C. 1961. Organization of the centroplasm in *Nostoc pruniforme. Iowa Acad. Sci.* **68:** 86-89.

Jensen, T.E., and Clark., R.L. 1969. Cell wall and coat of the developing akinete of a *Cylindrospermum* species. *J. Bacteriol.* **97:** 1494-1495.

Jost, M. 1965. Die Ultrastruktur von *Oscillatoria rubescens* DC. *Arch. Mikrobiol.* **50:** 211-245

Kale, S.R. and Talpasayi, E.R.S. 1969. Heterocyst : a review. *Indian Biologist,* **1:** 19-29.

Kaushik, M. and Kumar, H.D. 1970. The effect of light on growth and development of two nitrogen fixing blue-green algae. *Arch. Mikrobiol.* **74:** 52-57.

Kumar, H.D. and Gupta, M. 1979. Effect of cyclic nucleotides on morphogenesis in *Nostoc muscorum. Arch. Mikrobiol.* **119:** 183-186.

Ladha, J.K. and Kumar, H.D. 1975. Developmental stages in a non-heterocystous filamentous cyanophyte. *New Phytol.* **70:** 477-483.

Lang, N.J. 1965. Electron microscopic study of heterocyst development in *Anabaena azollae* Strasburger. *J. Phycol.* **1:** 127-134.

Lamont, H.C. 1969. Shear-orientated microfibrils in the mucilaginous investments of two motile oscillatorian blue-green algae. *J. Bacteriol.* **97:** 350-361.

Lawry, H. and Jensen, T.E. 1979. Depostion of condensed phosphate as an effect of varying sulphur deficiency in the cyanobacterium *Synechococcus* species (*Anacystis nidulans*). *Arch. Microbiol.* **120:** 1-7.

Lawry, N.H. and Simon,R.D. 1982. The normal and induced occurrence of cyanophycin inclusion bodies in several blue-green algae. *J. Phycol.* **18:** 391-399.

Lazaroff, N. 1966. Photoinduction and photoreversal of Nostocacean developmental cycle. *J. Phycol.* **2:** 7-17.

Lazaroff, N. and Vishniac, W.C. 1961. The effect of light on the developmental cycle of *Nostoc muscorum,* a filamentous blue-green alga. *J. gen. Microbiol.* **25:** 365-374.

Lex, M. and Carr, N.G. 1974. The metabolism of glucose by heterocysts and vegetative cells of *Anabaena cylindrica. Arch. Microbiol.* **101:** 161-167.

Miller, M.M. and Lang, N.J. 1968. The fine structure of akinete formation and germination in *Cylindrospermum. Arch. Mikrobiol.* **60:** 303-313.

Nichols, J.M. and Adams,D.G. Akinetes. In: Carr,N.G. and Whitton,B.A. (eds). The *Biology of Cyanobacteria*. Blackwell Scientific Publications, London, pp 387-412.

Peat,A. and Whitton, B.A. 1967. Environmental effects on the structure of the blue-green alga, *Chlorogloea fritschii. Arch. Mikrobiol:* **57:** 155-180.

Roussard-Jacquemin, M. 1983. Etude ultrastructure de la differentiation des heterocystes chez la cyanobacteria *Anabaena cylindrica* Lemm. *Can. J. Microbiol.* **29:** 1564-1575.

Reddy, P.M. and Talpasayi,E.R.S. 1974. Heterocyst formation in a blue-green alga, *Cylindrospermum. Nature,* **249:** 493-494.

Simon, R.D. 1980. DNA content of heterocysts and spores of the filaments cyanobacterium *Anabaena variabilis. FEMS Microbil. Lett.* **8:** 241-245.

Singh, H.N. and Srivastava,B.S. 1968. Studies on morphogenesis in a blue-green alga.I. Effect of inorganic nitrogen sources on developmental morphology of *Anabaena doliolum. Can. J.Microbiol.* **14:** 1341-1346.

Singh,H.N. and Sunita,K.M. 1974. A biochemical study of spore germination in the blue-green alga, *Anabaena doliolum. J. Exp.Bot.* **25:** 837-845.

Singh,R.N. and Tiwari,D.N. 1970. Frequent heterocyst germination in the blue-green alga, *Gloeotrichia ghosei* Singh. *J. Phycol.* **6:** 172-176.

Singh,R.N. and Viswanathan,I. 1972. Heterocyst differentiation in *Campylonema lahorense* Ghose. In: Desikachary,T.V. (ed) *Taxonomy and Biology of Blue-green algae.* Univ. Madras, pp 209-217.

Singh.R.N., Tiwari,D.N. and Singh,V.P. 1972. Genetic basis of cellular differentiation in Blue-green algae. In:Desikachary, T.V. (ed) *Taxonomy and Biology of Blue-green Algae.* Univ. Madras, pp 27-37.

Stewart, W.D.P., Rowell, P., Hawkesford,M., Sampaio,M.J.A.M., Ernst,A. 1982. Nitrogenase and aspects of its regulation in cyanobacteria. *Israel J. Bot.* **31:** 168-189.

Sutherland, J.M., Herdman,M. and Stewart,W.D.P., 1979. Akinetes of cyanobacterium *Nostoc PCC* 7524: macromolecular composition, structure and control of differentiation. *J. gen. Microbiol.* **115:** 273-287.

Thuret, G. 1875. Essai de classification des Nostochinees. *Ann. Sci. nat. Bot.* ser 6. **1:** 372-382.

Walsby, A.E. and Nichols,B.W. 1969. Lipid composition of heterocysts. *Nature,* **221:** 673-674.

Wilden,D.C., and Mercer, F.V. 1963. The Ultrastructure of the heterocyst and akinete of blue-green algae. *Aust. J.biol. Sci.* **16:** 585-586.

Wolk,C.P., and Simon,R.D. 1969. Pigments and lipids of heterocysts. *Planta,* **86: 92-97.**

Ind. Phyco. Rev. 1993.
2: 41-58.

HABITATS AND DISTRIBUTION

A.S. AHLUWALIA
Phycology Laboratory,
Botany Department, Panjab University,
Chandigarh - 160 014, INDIA

INTRODUCTION

In the continuum of life from the very small prion particle through the unicellular procaryotic and eucaryotic microorganisms and finally to higher plants and animals, the cyanophycean (blue-green algal) members had been occupying a unique position, although in recent years some microbiologists have put these procaryotic microorganisms alongwith phototrophic bacteria (Stanier *et al.*, 1971, 1978). Cyanophyceae forms the oldest group of algae with definite fossil remains in the form of stromatolites dating back to almost 3000 million years (Lee, 1989). Evolution of these organisms eventually built up the oxygen content of the atmosphere.

The blue-green algae are widely distributed in practically all types of habitats in temperate, tropical and polar regions, which can support life. Most commonly, these occur in fresh water, marine or brackish water, however, these can also grow in hot springs, deserts, saline lakes, the antarctic, oil field ponds to lichen associations on bare rocks. Professor Fritsch had rightly stated that the Myxophyceae are a highly successful group and enjoy a wide distribution in all kinds of habitats (Fritsch, 1945). As far as the diversity of their habitats and distributional pattern is concerned, blue-green algae come next to bacteria. The secret of their wide range of distribution lies in their photoautotrophic nature with an ability to fix dinitrogen of the atmosphere and use it for their growth and development (Carr and Whitton, 1982). In addition, blue-green algae can fulfil their carbon requirement by utilizing the carbondioxide of the ambient atmosphere or from carbonates/bicarbonates present in water.

These organisms are of considerable ecological importance because of their well known function as the primary producers in the food chain and their capability of increasing the fertility of terrestrial and aquatic habitats. Their role in colonization of newly exposed land surfaces and reclamation of alkaline (usar) soils has been known since long (Singh, 1961; Holm-Hansen, 1968). There can be composite growths of diverse blue-green algae forming frequent associations and in these the colonial forms are often the precursors of filamentous types in the colonization of terrestrial and aquatic habitats. The role of plasmids in natural distribution of blue-green algal species is as yet unknown. Blue-green algae are autotrophic organisms though a few are capable of heterotrophic nutrition. Occurrence of some blue-green species in polluted environs may give an indication of their capacity to grow in oxygen-deficient habitats. Although some algal forms can survive in dark for long periods, but their growth increments are mainly dependent upon the period of illumination. These algae usually grow profusely in shallow older eutrophic lakes to form water-blooms which sometimes result in serious losses of economic and asthetic resources.

Nature has been extremely lavish in the humid tropics and subtropics for supporting an abundant flora of blue-green algae throughout the year. In India, this flora is generally active only during the rainy season because various substrata inhabited by these algae become moist for quite sometime to initiate their luxuriant growth (Singh, 1961). Credit for the first convincing demonstration concerning the distribution of blue-green algae inhabiting Indian rice fields and their importance as biofertilizers goes to Professor De (De, 1939). Blue-green algae comprise a wide assemblage of free living and symbiotic forms where their ecological importance is of significance. The seasonal variations in their appearance quite often lead to problems in evaluating their distributional pattern.

The identification of some blue-green algae is difficult because of their pleiomorphic nature and uncertainties in taxonomy. The taxonomy of this group has recently been undergoing many changes (Friedmann and Borowitzka, 1982). So species names are often used in different meanings by different authors. In the following pages, an attempt has been made to summarize the information available on the habitats and distribution of this important group of algae.

The presence of cyanophytes have been reported from mangroves (Cordero, 1978). However, no report has yet been made of blue-green algae in the open sea, although this phenomenon is more or less universal (Fogg, 1982).

Members of blue-green algae may be distributed as free living organisms or by forming symbiotic associations with some members of Fungi, Bryophytes, Pteridophytes, Gymnosperms, Angiosperms etc. Most of the forms are exclusively free-living. These may be distributed in various aquatic habitats, thermal environs, snow-clad mountains and terrestrial habitats.

A. **Free living blue-green algae :**

1. Aquatic blue-green algae
2. Terrestrial blue-green algae
3. Blue-green algae of specialized habitats

1. *Aquatic blue-green algae :*

These include mainly the freshwater and marine blue-green algae.

(a) *Freshwater algae* : Most of the aquatic blue-green algae are cosmopolitan in their distribution though certain heterocystous branched forms are largely confined to warmer regions. The freshwater Cyanophyceae are important as lithophytes and epiphytes. Blue-green algae are widely distributed in the benthos and plankton of fresh waters of all kinds. *Chamaesiphon* is widely distributed in fresh waters whereas *Nostochopsis lobatus* is commonly found in stagnant waters (Borzi, 1917). The gelatinous mass formed by *Nostochopsis* can be attached or free floating.

Some species of *Pleurocapsa* are found in well aerated habitats as littoral regions of lakes and rapid streams. The tubular colonies of *Wollea bharadwajae* were reported to be free floating in freshwater bodies (Singh, 1942). *Scytonema ocellatum* was reported from a freshwater pond for the first time in India from Andaman Island (Prasad and Srivastava, 1984). Many of the Stigonematalian members occur in freshwater habitats. Several freshwater forms also inhabit bottom of shallow and deeper water and represent a significant part of the plankton.

Planktonic forms usually have pseudovacuoles and hence participate in the formation of water-blooms (Palmer, 1980). Blooms are normally common in summer than in winter. Presence of rich organic matter in a water body (eutrophication) accelerates the formation of water-blooms (Table 1). The bottom of stagnant ponds are often covered as thin films by the members of Oscillatoriaceae while spherical balls are formed due to Chroococcales. Some lithophytic forms belonging to Chamaesiphonales, Pleurocapsales and *Phormidium* are characteristically distributed in fast-flowing streams. These forms are normally responsible in imparting different colours to the underwater rocky surfaces.

Table 1 : Blue-green algae common in organically enriched areas.

Agmenellum quadruplicatum
Anabaena constricta
Arthrospira jenneri
Oscillatoria chalybea
O. chlorina
O. formosa
O. lauterbornii
O. limosa
O. princeps
O. putrida
O. splendida
O. tenuis
Phormidium autumnale
P. uncinatum

Some filamentous members of Oscillatoriaceae behave as planktonic forms. These include *Lyngbya contorta* and *Spirulina* species. Among heterocystous forms, *Aphanizomenon* occurs as a planktonic form (Rose, 1934). Blooms of blue-green algae occur sporadically throughout the summer months with the decline in one species often followed by an increase in another. During calm weather, the algal colonies may rise to the surface forming a thick scum or dense clamps. The bloom forming genera in temperate water normally include *Oscillatoria, Lyngbya, Microcystis, Aphanizomenon,*

Anabaena and *Gloeotrichia.* Various members of the genera *Anabaena, Anabaenopsis, Raphidiopsis, Cylindrospermum* and *Wollea* have been found in many of the inland waters (Singh, 1961).

In the Samber salt lake (Rajasthan), blue-green algae like *Anabaenopsis arnoldii* and *Arthrospira platensis* grow abundantly and sometimes even interfere with the salt production.

Vertical movements have been observed for planktonic *Oscillatoria, Raphidiopsis* and *Arthrospira*. The algal blooms are much common in the temple tanks in India (Venkataraman *et al.,* 1974). The vertical migration is associated with the formation and collapsing of gas vacuoles within the algal cells (Walsby, 1972). Although some of the bloom forming blue-green algae are known to be toxic to animals but this aspect has not got much attention in India, may be because of the low occurrence of such toxic effects here. In India, aegagrophilous balls of *Schizothrix lacustris* have been reported in shallow waters (Desikachary and Ramakrishnan, 1953; Desikachary, 1959).

Kumar (1990) has discussed the Indian work on freshwater algae. Many of the Indian rivers have been studied for phytoplankton and periphyton communities. In the Kashmir State, ephemeral thermal stratification was earlier shown in oligotrophic lakes supporting poor growth of phytoplankton. Later, however, many of these lakes are becoming eutrophicated and show maximum growth of blue-green algae during the month of October.

The range of temperatures as well as light intensity in Indian plains, is normally condusive to good growth of blue-green algae. The pH value of most Indian rivers remains on the alkaline side thus supporting the blue-green algal growth (Prasad and Singh, 1980). In general, the density of total algae in rivers is the highest in August-October and February-March, and is the lowest in May to July.

Blue-green algae are characteristic of waters receiving high nutrient inputs, yet the maximum algal biomass occurs in low levels of dissolved nutrients. The blue-green algae have a high affinity for phosphorus and can outnumber other algae at low nutrient concentrations (Shapiro, 1973).

Many freshwater blue-greens have the capacity to precipitate lime which may lead to extensive deposits of lime. Species of *Rivularia, Schizothrix, Plectonema* and *Aphanothece* are mainly involved in such deposition alongwith some thermal algae.

(b) *Marine blue-green algae :* In addition to distribution in freshwater habitats, blue-green algae with the exception of Stigonematalean members form an important element in marine vegetation. These marine forms constitute an important inhabitation in the marine intertidal and supratidal belt, while a few are planktonic as well (Desikachary, 1959). Many red pigment (Phycoerythrin) containing algal forms owe their distribution to deep regions in the sea due to their ability of chromatic adaptation. Such a capacity of adjusting to different light regimes indicate a better utilization of the available light and this may be partially responsible for their ubiquitous occurrence in almost all kinds of aquatic habitats (Ahluwalia *et al.*, 1980).

Members of Chroococcaceae, Oscillatoriaceae and Rivulariaceae are well represented in the littoral zone, forming conspicuous growth belts on the rocks. Species of *Chlorogloea* and *Entophysalis granulosa* are distributed chiefly in the spray zone of the sea-shore. *Dermocarpa* is largely a marine epiphytic alga. Many species of *Calothrix, Scytonema* and *Brachytrichia* are distributed as the marine littoral community. Other examples include *Aphanocapsa littoralis, Gloeocapsa crepidinum, Rivularia atra, Lyngbya majuscula* etc. Members of Oscillatoriaceae also occur in shallow water.

Trichodesmium forms extensive bloom in the Indian seas particularly the Arabian sea. In the west coast of India, *Lyngbya, Pleurocapsa and Calothrix* constitute the common blackish green belt in the supralittoral zone with an average vertical expanse of 6-12 meters from the high water mark (Misra, 1960; Srinivasan, 1960). In such habitats *Lyngbya majuscula* has been dominating over *Phormidium corium, Pleurocapsa fuliginosa* and *Calothrix contarenii*. The distribution of Cyanophycean members is very much restricted in the east coast.

The commonly occurring blue-greens in brackish waters near the sea include *Trichodesmium, Lyngbya, Oscillatoria, Microcoleus* and *Chroococcus*. In India, more than 100 species are known from saline habitats including a major share of marine forms. Members belonging to Chamaesiphonales and Pleurocapsales form the major constituent of marine forms alongwith few species of other orders. Anand and Venkatesan (1985) described some blue-greens from the salt pans at Covelong (Madras). About 25 taxa of blue-green algae have been reported from backwaters in Muttukad, estuaries of Adyar and salt marshes of Pichavarm. The pattern of their distribution showed the dominance of nonheterocystous filamentous members in all the sampling sites (Anand *et al.*, 1986).

2. *Terrestrial Blue-green algae :*

Terrestrial blue-green algae occur worldwide and include the most hostile environment like extreme arid cold and hot deserts where they often are the only or major primary producers. A number of terrestrial niches are occupied by such algaeoccurring either on surface or at a depth of upto several centimeters in soil. In addition, they survive in and on rocks and in caves, a habitat with very low light intensity. Furthermore, they inhabit permanent snow and ice-fields and even animals and plants. In many of such habitats, frequent fluctuations of environmental conditions occur, in sharp contrast to the mostly stable aquatic habitats. The species of *Stigonema* are generally subaerial. Most of the Chroococcalean species and many forms belonging to other orders of blue-green algae are mainly terrestrial. Some filamentous species (Oscillatoriaceae) commonly form wefts on damp soil, especially in the warmer parts of the earth. Few members belonging to Stigonematales are distributed in terrestrial habitats, e.g. many species of *Fischerella* have also been recorded to be terrestrial (Singh, 1961).

(a) *Soil blue-green algae* : Soil is the best studied terrestrial algal habitat. Estimates of algal biomass in soil populations differ greatly among various researchers as no standards exist for these methods. The most important factors controlling algal populations in soil include light, humidity, temperature, availability of nutrients and pH. Density of these algae is often the highest in upper layers of soil and fall off rapidly with depth (Peterson, 1935). The algae surviving below the level of light penetration may be because of chemoheterotrophy (Khoja and Whitton, 1975). Algae growing completely exposed on the soil surface are called saphophytes while those growing below the surface are cryptophytes.

Many soils are subjected to fluctuations in the level of humidity and soil algae are generally adapted to survive periods of desiccation. Viable soil alga, *Nostoc commune* and others have been recovered from 25 to 87 years old herbarium specimens (Bristol, 1919; Trainor, 1985). Some desiccation resistant soil algae form resting spores (akinetes), e.g. *Anabaena, Cylindrospermum, Nostoc, Aulosira* and *Nodularia*. Blue-green algae, especially the filamentous forms, generally prefer neutral and alkaline soils. Species of *Microcoleus* and *Schizothrix* often form crusts on soils (Durrell and Shields, 1961). The heterocystous algal forms are of great significance to soil ecosystem.

Species of *Nostoc, Anabaena, Calothrix* and *Cylindrospermum* are often common on the soil surface soon after the rains. Such algae can have importance in the protection of eroded soil. Although sheaths seem to be playing a role in the survival of algae under dry conditions, but participation of protoplast in this process cannot be ruled out as some forms without well developed sheaths do survive in such habitats.

A form of *Aulosira fertilissima* was reported to grow as a blanket in the lowland and usarland rice fields in Allahabad (Varma *et al.*, 1964). Raju (1964) reported *Aulosira terrestre* alongwith *Scytonema hofmanni, Nostoc corneum* and *Anabaena* species from the uncultivated soils of Hyderabad. About 19 species of Cyanophycean algae were described from Jodhpur and its environs (Goyal, 1964; Gupta, 1972) whereas Gupte (1964) reported 23 forms of this group from Ajee river near Rajkot. *Chlorogloea fritschii* was first described by Mitra (1950) from cultivated and uncultivated soils near Allahabad. A good number of blue-green algal forms has been recorded from Bhopal especially during the period from September to November (Bendre and Agarker, 1965). Chaporkar and Gangawane (1984) reported 33 Cyanophycean forms from various crop field soils of Marathwada (Maharashtra). In a study, Goyal *et al.* (1984) found that algal flora of rice field soils of Jammu division harbour more blue-green algae than Srinagar division.

The soil blue-green algae have been extensively studied mainly for their distribution in rice field soils (Singh and Singh, 1989). Most of the cultivated soils and grasslands support an abundant blue-green algal flora (Table 2). In India, this flora shows a distinct periodicity during and just after monsoon (Singh, 1961). Pandey (1966) found that about 70% of the algal species in paddy soils of Uttar Pradesh were blue-green algae. Similarly, Aiyer (1965) studied the algal flora of rice fields of seven districts of Kerala and found that blue-green algae constitute 20-76% and dominate other groups of algae. It was found that *Aulosira fertilissima* commonly encountered in the rice field soils of Bihar and Uttar Pradesh as a dominant form, was absent in the paddy soils of some parts of south India (Khurana and Venkataraman, 1968).

Table 2.: Season distribution of indigenous nitrogen-fixing blue-green algae in an unfertilized rice field. (Annual reports of the All India Coordinated Project on Algae, 1976-1981, DST, New Delhi).

Month	Algal species
January-February	*Nostoc muscorum* (C)
March	*Nostoc muscorum* (C), *N. punctiforme* (A), *Anabaena variabilis* (C), *A. ambigua* (C)
April-May	*Nostoc muscorum* (A), *N. punctiforme* (A), *Anabaena variabilis* (A), *A. ambigua* (C)
June	*Nostoc muscorum* (A), *N. punctiforme* (A), *Anabaena variabilis* (A), *A. ambigua* (A), *Aulosira fertilissima* (R)
July	All the species during June as (A).
August	All the above plus *Cylindrospermum muscicola* (C)
September	All the above plus *Westiella* sp.(C)
October	All the above (A)
November	*Nostoc punctiforme* (C), *Westiella* sp. (C)
December	*Nostoc muscorum* (C)

A = Abundant, C = Common, R = Rare

Arora (1985) studied the distribution of blue-green algae in Punjab and reported the isolation of ten algal genera. *Anabaena* and *Calothrix* species have been observed to the more common in their distribution as these are relatively more thermotolerant than others (Arora *et al.*, 1989; Ahluwalia *et al.*, 1992).

Several nitrogen-fixing blue-green algal species have been isolated from temperate soils (Bortels, 1940). In desert soils, the blue-green algae are often the dominant organisms in the algal crusts (Fletcher and Martin, 1948). *Nostoc commune* is widespread on sandy soils in Asia. Such forms are known as epipsammic algae.

In alkaline (Usar) soils the growth of blue-green algae can be encouraged by making shallow ponds in the rainy season (Singh, 1950). These algae are also very important ecologically as pioneer forms. Species of *Chroococcus* and *Microcoleus* are frequent pioneers on the mud of salt-marshes (Carter, 1933). Algal forms growing on mud are called epipelic. The distribution of blue-green algae in salt marshes and other saline habitats is due to their capacity to withstand marked changes in the concentration of the medium. Halophilic species can grow and multiply in solutions more concentrated than 3 molar sodium chloride, e.g. *Aphanocapsa littoralis, Spirulina subsalsa* and *Phormidium tenue.*

(b) *Lithophytic blue-green algae :* These organisms live on or within rock substrates. Several blue-green algae are important as lime-boring forms and participate in the deposition of calcareous and siliceous rock-masses. These can be found on the rocks of rushing torrents and cataracts (Fritsch, 1945). Members of Chroococcales showing lithophytic nature were put under Entophysalidaceae (Geitler, 1932). Species of *Chamaesiphon* constitute an important part of the lithophytic flora of streams (Fritsch, 1929) whereas species of *Pleurocapsa* are widely distributed as marine lithophytes inhabiting the upper part of the shore.

Rock inhabiting algae are eurytherm as their environment undergoes great daily and annual temperature variations. In temperate regions the surface temperature can go upto 50^0 C in summer while in winter the algae have to survive subzero temperature. In tropical areas, temperature always remains above the freezing point. In polar regions, algae are frozen for most of the year.

The inorganic nutrients necessary for algal growth are probably found in the soil for hypolithic algae and in rock for other lithophytic algae. In part, they may even be conveyed to rocks by rain, snow and dry atmospheric fall out.

Golubic *et al.* (1981) has proposed following terminology for different lithophytic habitats:

i) Epilithic blue-green algae : These forms colonize the external exposed surface of rocks as a crust or as a thick layer. These are generally the first photosynthetic organisms that colonize these surfaces and are the only ones that can survive there (Fritsch, 1907). Light intensity and humidity are the important factors that determine the composition of an epilithic algal vegetation (Hoffmann, 1989). The subaerial growth on calcareous rocks has a good number of blue-green algae including *Gloeocapsa alpina*-forming black crust, *Scytonema myochrous, Calothrix parietina, Aphanocapsa* and *Chroococcus.* At high light intensities, *Tolypothrix byssoidea* replaces *Scytonema.* In areas with a high degree of humidity, many Chroococcalean and filamentous members produce leathery outgrowths. Extensive growth of subaerial algae occurs in warmer areas having high atmospheric humidity. The landscape in such habitats gets covered by the various growth forms of blue-green algae, probably representing different stages in the colonisation of the substratum (Fritsch, 1945).

Marathe and Chaudhari (1975) reported considerable rock corrosion under epilithic algal mats. Some epilithic algae (*Pleurocapsa, Plectonema gloeophilum*) can participate in rock formation by depositing carbonate crusts on rock surfaces. This crust can protect these algae from high light intensities and help in retaining humidity.

A number of filamentous nonheterocystous blue-green algae usually form the dominant flora on the walls of buildings. Species of *Lyngbya* and *Plectonema* have been reported to be the most frequent and abundant algae from the walls of ancient Roman frescoes which seem to be the primary agent of their biodeterioration (Grilli Caiola *et al.*, 1987). Heterocystous forms are normally absent or poorly distributed in such habitats. Chadha and Pandey (1982) found various blue-green algae from the walls of buildings in and around Allahabad University Campus. Indications of variations in their distributional pattern on different types of wall bases were also reported. *Gloeocapsa polydermatica* was reported to be frequent on plastered white washed and stone walls while it was not recorded on cemented and brick walls. In addition, the occurrence of *Chroococcus minutus, C. pallidus, Gloeocapsa granosa, Aphanocapsa biformis, A. grevillei* and *Aphanothece pallida* was reported. The filamentous forms comprised *Phormidium corium, Lyngbya sordida, Oscillatoria curviceps, Microcoleus* sp., *Scytonema ocellatum, S. guyanense* and *Calothrix castelli*. It was interesting that none of these algal forms has exclusively wall habitat (Chadha and Pandey, 1982). The survival of these algae on such unhospitable habitats, in part, seems related to the presence of sulphydryls (Tripathi and Talpasayi, 1980).

ii) Hypolithic blue-green algae : Such algae live under pebbles and small stones lying or buried in the soil. These are generally known from deserts and semideserts. The algal vegetation consists of a thin layer of blue-green algae growing under translucent pebbles from near the soil surface down to about 4 cm. The vegetation is mainly formed by coccoid and filamentous blue-green algae, e.g. *Chroococcidiopsis* sp., *Aphanocapsa* sp., *Aphanothece* sp., *Gloeocapsa* sp., *Lyngbya* sp., *Microcoleus chthonoplastes, Myxosarcina* sp., *Nostoc* spp., *Plectonema* sp., *Schizothrix calcicola, Scytonema* sp., and *Tolypothrix byssoidea*. Broady (1981) lists species of *Aphanothece, Chroococcidiopsis, Lyngbya, Plectonema, Calothrix, Nodularia, Nostoc* and *Tolypothrix* from Antarctica.

iii) Euendolithic blue-green algae: These algae colonize the interior of rocks and actively penetrate the rock surface like the boring organisms. Limestone boring algae have been known in freshwater and marine habitats. Only blue-greens have been known to occur as euendolithic terrestrial algae (Hoffmann, 1989). Danin (1983) found that unicellular blue-green algae can penetrate the rock @ 5 μm/year. Limestone boring algae include *Gloeocapsa* spp., *Scytonema* sp., *Chroococcus* sp., *C. lithophilus, Aphanocapsa endolithica* and *Schizothrix corciacea. Hyella* inhabits diverse calcareous substrata. In addition, such blue-green algae are also found in other calcareous substrata including shells of Molluscs, corals and calcareous deposition on some red algae. After boring into such structures, these algae make the rock surface soft, porous and erosive. The algae inhabiting shells of various animals are normally very indifferent to changes in salinity and temporary desiccation and may extend to a depth of about 50 meters.

iv) Chasmoendolithic blue-green algae : Such algae occur mostly in deserts and alpine regions. The debris carried by air accumulates in cracks and provide a chasmoendolithic habitat similar to hypolithic one. Therefore, this habitat have many common species to the latter (Broady, 1981). These algae occupy spaces within rocks that are open to the surface through cracks and fissures.

v) Cryptoendolithic blue-green algae : These algae colonize structural cavities within porous rocks at a depth of one to several millimeters. The rocks usually occupied by these algae are limestone and sandstone. The algal zone runs at a uniform depth following contours of the rock surface. Algae inhabiting such rocks in hot deserts are usually unicellular ones, e.g. *Chroococcidiopsis* sp., *Chroococcus turgidus.* In semideserts, few more forms including the filamentous ones, do occur.

B. Cave blue-green algae :

Such algae cover surface of cave walls with a dense cover, extending from entrance of caves to end of the photic zone. Temperature and humidity are generally even throughout the year but with a light gradient. This gradient controls the composition of these blue-green algae. Limestone cave walls have calcifying filamentous blue-green algae. *Scytonema julianum* occupies position near the entrance. Other forms include *Gloeocapsa alpina, Tolypothrix rupestris, Nostoc microscopicum, Chroococcus varius, Chlorogloea microcystoides and Aphanocapsa fuscolutea* (Hoffman, 1989).

3. *Blue-green algae of Specialized Habitats :*

a) Cryophytic blue-green algae : Blue-green algae are capable of surviving at both the temperature extremes though these are relatively poorly represented in the snow-fields. These algae inhabit frozen water at ice-air interface and are considered terrestrial in contrast to marine cryobiontic algae which occupy ice-water interface and are aquatic. Terrestrial cryobiontic algae occupy the spaces between ice crystals in the surface layers of snow and ice. They are visible only when aggregate in large numbers leading to coloration of ice. Such algae are active only for a limited period of the year. Blue-snow is caused by *Dactylococcopsis caucasica* and *Gloeothece transsylvanica.* Blue-green algae contribute nearly 26% of the total algal species recorded from snow and ice. Some blue-green algae *(Phormidium* spp.) flourish even in the frigid lakes of Antarctic.

b) Epiphytic blue-green algae : Epiphytic blue-green algae may be epiphloeic (occurring on tree bark) or epiphyllic (on leaves). This habitat is more arid than the

soil and is subjected to periodic desiccation. Bark fissures create more favourable microclimate for algal growth than other plant parts and the bark itself supplies nutrients for these algae. In tropical regions, blue-greens form a major part of epiphytic vegetation. Common epiphloeic forms are species of *Aphanocapsa, Gloeocapsa, Microcoleus, Schizothrix, Symploca, Tolypothrix, Scytonema, Porphyrosiphon, Nostoc* and *Fischerella* whereas *Scytonema javanicum, S. hofmanni, Stigonema hormoides* and *S. minutum* constitute the epiphyllous forms (Fremy, 1930). *Pulvinularia* has been recorded to occur as an epiphyte on *Fontinalis* (Borzi, 1916).

c) Epizoic blue-green algae : Some blue-green algae do occur as epizoic (on animals) or endozoic (inside animals). The genera present on hairs of sloths are *Oscillatoria, Nostoc, Fischerella, Chroococcus* and *Pleurocapsa* (Thompson, 1972). A unicellular blue-green alga living inside the hairs of polar bears in zoos has been recorded (Lewin and Robinson, 1979). Although algae living on animals are infrequent, still these animals may be important in the dispersal of these algae. *Loriella* was observed on human skull (Borzi, 1916). Some algae are found in associations with animals, e.g. gut of milkfish (Vicencio, 1977). *Lyngbya thornensis* was reported to grow on the antennae of *Daphnia pulex* (Bunting and Lund, 1956).

d) Thermophilic blue-green algae : Few blue-green algae have been recorded as the regular constituent of the flora of hot water springs. Typical examples include *Mastigocladus laminosus, Hapalosiphon laminosus, Phormidium laminosum, P. valderianum, Oscillatoria tenuis* and *Synechococcus thermalis* (Prasad and Srivastava, 1965; Thomas and Gonzalves, 1965 and Vashishta, 1966). Some of the thermophilic algae play a role in the formation of rock-masses. The highest temperature at which blue-green algae have been recorded varied from 70-75^{0}C and no algal form was observed around 94^{0}C (Vashishta, 1965, 1966). Elenkin (1914) reported algal life at an upper limit of 85^{0}C. Many of the thermal blue-greens have been recorded from non-thermal (normal temperature) waters as well. The thermal algae are usually categorized on their tolerance to different temperatures (Table 3).

Table 3 : Distribution of thermophilic blue-green algae at different thermal environs in India.

Temperature	Abundant species
40-45^{0}C	*Anabaena fertilissima, Chroococcus minor, C. minutus* var. *thermalis, C. hansgirgi* var. *minor, Synechocystis manikaranensis, S. minuscula, Gloeocapsa pleurocapsoides, Oscillatoria subbrevis, O. jasorvensis, O. tenuis, O. formosa, O.*

Temperature	Abundant species
	proboscidea, O. sancta, Phormidium acidophilum, P. valderianum, Lyngbya nigra, L. martensiana, Microcoleus apiculatus.
46-50^{0}C	*Chroococcus yellowstonensis, Synechococcus elongatus, S. amethystinus, S. lividus, Myxosarcina amethystina* var. *thermalis, Oscillatoria angustissima, O. geminata* var. *fragilis, Phormidium tenue, P. valderianum.*
51-59^{0}C	*Phormidium laminosum, Mastigocladus laminosus.*
60-65^{0}C	*Synechococcus arcuatus, Oscillatoria filiformis.*
66-70^{0}C	*Synechocystis thermalis, Synechococcus minimus, S.vulcanus.*
Above 70^{0}C	*Chroococcus turgidus* var. *thermalis, Phormidium bijahensis, Synechococcus eximius, S. vulcanus* var. *bacillarioides, Westiella goshii.*

C. Symbiotic blue-green algae .

Myxophycean microenvironment is influenced by bacteria, fungi, lichens, higher algae and mosses, which often accompany the blue-green algae in their natural habitats (Mague, 1977). Therefore, number of blue-green algae live in association with other organisms. Basically blue-green algae occur in two types of associations based on the position of the alga as extracellular or intracellular (Table 4). Blue-green algae represent the algal part (phycobiont) in about 8% of the lichen species, as symbionts in the cells of diverse unicellular animals including some colourless organisms and plants belonging to different groups.

Table 4 : Major symbiotic associations formed by different blue-green algae.

Plant and Group	Location of alga in the plant
Rhizosolenia : Alga	Intracellular
Peltigera : Lichens	Extracellular (continuous or in cephalodia)
Blasia : Bryophyte	Extracellular (leaf space)
Azolla : Pteridophyte	Extracellular (leaf space)
Cycas : Gymnosperm	Extracellular (root cortex)
Gunnera : Angiosperm	Intracellular (stem gland)

Various species of blue-green algae (*Scytonema, Gloeocapsa, Nostoc, Anabaena, Stigonema*) have been found as endophytes or symbionts occurring in association with various plants. These include fungi, liverworts, water ferns, Gymnosperms and Angiosperms. In *Gunnera,* the alga occurs within three wart-like swellings on the leaf bases (stem glands). The algal cells occupy the cells in *Gunnera* while in others they inhabit intercellular spaces. The algal zone in *Cycas* lies in the cortex of coralloid roots whereas the alga inhabits leaf cavities in fronds of *Azolla* (Rai, 1990).

Nitrogen-fixing symbiosis of diatom *(Rhizosolenia)* and blue-green alga (*Richelia*) has been reported to bloom extensively (Heinbokel, 1986; Villareal, 1989). Marine sponges also contain symbiotic blue-green algae and the endosymbiont is either free living in the mesophyll or contained in the digestive vacuoles (Berthold *et al.,* 1982). It can easily be deduced from the above discussion that blue-green algae occur in most of the environments on earth except water and soil with a much higher acidic nature.

ACKNOWLEDGEMENTS

I thank Dr. (Mrs.) Kanwaljit Kaur Ahluwalia for her constant inspiration during the course of this work.

REFERENCES

Ahluwalia, A.S., R.K.Rai and H.D. Kumar, 1980. Chromatic adaptation and photoreversal in a blue-green alga *Calothrix clavata* West. *J.Biosci.* **2:** 63-68.

Ahluwalia, A.S., S.Dahuja and Reena, 1990. Role of Blue-green algae in increasing the growth and yield of rice (*Oryza sativa* var. PR 106). In :"Proc. Natl. Symp. Cyanobacterial Nitrogen Fixation", (Kaushik, B.D. ed.).Associated Publish. Co., New Delhi, p. 307-313.

Aiyer,R.S. 1965. Comparative algological studies in rice fields in Kerala State. *Agri. Res. J. Kerala* **3:** 100-105.

Anand,N., E. Mohan, R.S.S. Hopper and T.D. Subramanian, 1986. Taxonomic studies on blue-green algae from certain marine environments. *Seaweed Res. Utiln.* **9:** 49-56.

Anand, N. and N. Venkatesan, 1985. Note on blue-green algae from salt pans. *Seaweed Res. Utiln.* **7:** 101-103

Arora,J.K., 1985. Role of blue-green algae in nitrogen fixation in relation to rice crop. *M.Sc. Thesis,* Punjab Agric. Univ., Ludhiana.

Arora,J.K., H.S. Garcha, M.S. Pandher and R.P. Gupta, 1989. Evaluation of the most prevalent blue-green algae for nitrogen-fixing capacity and thermotolerance in Punjab. *Ann. Biol.* **5:** 77-80.

Bendre,A.M. and M.S. Agarker, 1965. Myxophyceae of Bhopal and its environs. *Phykos* **4:** 76-82.

Berthold,R.J., M.A. Borowitzka and M.A. Mackay, 1982. The ultrastructure of *Oscillatoria spongeliae,* the blue-green algal endosymbiont of the sponge *Dysidea herbacea. Phycologia* **21:** 327-335.

Bortels,H., 1940. Uber die Bedeutung des Molybdans fur Stickstoffbindende Nostocacean. *Arch. Mikrobiol.* **11:** 155-186.

Borzi, A., 1916. Studi sulle Mixoficee-II. *Nuovo Giorn. Bot Ital,* **23:** 559-588.

Borzi,A., 1917. Studi sulle Mixoficee-III. *Nuovo Giorn. Bot Ital.,* **24:** 17.

Bristol,B.M. 1919. On the retention of vitality by algae from old stored soils. *New Phytol.* **18:** 92-107.

Broady,P.A., 1981. The ecology of sublithic terrestrial algae at the Vestfold Hills, Antarctica. *Brit. Phycol. J.* **16:** 231-240.

Bunting,W. and J.W.G. Lund, 1956. A new blue-green alga epizoic on *Daphnia dulex* L. *Naturalist,* Hull No. 858, p. 88-90.

Carr, N.G. and B.A. Whitton (eds), 1982. *The Biology of Cyanobacteria.* Blackwell Scientific Publi., Oxford. 688 p.

Carter,N., 1933. A comparative study of the algal flora of two salt marshes. *J. Ecol.* **21:** 128-208.

Chadha,A. and D.C. Pandey, 1982. Distribution of algae on the walls of buildings. *Phykos* **21:** 122-123.

Chaporkar, C.B. and Gangawane, L.V., 1984. Blue-green algae of some cultivated soils of Marathwada, Maharashtra-I. *Phykos* **23:** 55-58.

Cordero, P.A., Jr., 1978. Phycological observations VI. Mangrove associated algae of Alkan, Philipines. *Kalikasan Phil.J.Biol.,* **7:** 275-296.

Danin, A., 1983. Weathering of limestone in Jerusalem by Cyanobacteria. *Z. Geomorphol. N.F.* **27**: 413-421.

De, P.K., 1939. The role of the blue-green algae in nitrogen fixation in rice fields. *Proc. Royal Soc. London* B **127**: 121-139.

Desikachary, T.V., 1959. Cyanophyta. Indian Council of Agricultural Research, New Delhi, 686 p.

Desikachary, T.V. and K.R. Ramakrishnan, 1953. On an interesting aegagrophilous *Schizothrix* from Madras Province forming balls. *Proc. Indian Sci. Congr.*, Part III.

Durrell, L.W. and L.M. Shields, 1961. Characteristics of soil algae relating to crust formation. *Trans. Amer. Microscop. Soc.* **80**: 73-79.

Elenkin, A.A., 1914. Uber die Thermophilen Algen-formationen. *Bull. Jardin. Imp. Bot. Pierrela Grand.* **14**: 62-110.

Fletcher, J.E. and W.P. Martin, 1948. *Ecology* 29 : 95 (Quoted from Singh, 1961).

Fogg, G.E. 1982. *Marine plankton.* In : "The Biology of Cyanobacteria". (Carr., N.G. and B.A. Whitton, eds.) Blackwell Sci. Pub., London. p. 491-513.

Fremy, P., 1930. Les Myxophycees de l'Afrique Equatoriale Francaise. *Arch. Bot Mem.* **3**: 1-508.

Friedmann, E.I. and L.J. Borowitzka, 1982. The symposium on taxonomic concepts in blue-green algae. Towards a compromise with the bacteriological code. *Taxon* **31**: 673-683.

Fritsch, F.E., 1907. A general consideration of the subaerial and freshwater algae of Ceylon. *Proc. Royal Soc. London* B **79**: 197-254.

Fritsch, F.E., 1929. The encrusting algal communities of certain fast flowing streams. *New Phytol.* **28**: 165-196.

Fritsch, F.E., 1945. *The Structure and Reproduction of the Algae.* Vol. 2. Cambridge University Press, Cambridge 939 p.

Geitler, L., 1932. *Cyanophyceae.* In: "Krytogamenflora von Deutschland, Osterreich and der Schweiz" (Rabenhorst, L., ed.) Vol. 14. Akademische Verlagsgesellschaft, Leipzing.

Golubic,S., I. Friedman and J. Schneider, 1981. The lithobiontic ecological niche, with special reference to microorganisms. *J. Sediment. Petrol.* 51: 475-478.

Goyal, S.K., 1964. Algal flora of Jodhpur and its environs -II. Cyanophyta. *J. Bombay Nat. Hist. Soc.* **61**: 69-73.

Goyal, S.K., B.M. Sharma and R.S. Gupta, 1984. Algae flora of rice field soils of Jammu and Kashmir State. *Phykos* **23**: 59-64.

Grilli Caiola, M., C. Forni and P. Albertano, 1987. Characterization of the algal flora growing on ancient Roman frescoes. *Phycologia* **26**: 387-390.

Gupta,R.S., 1972. Blue-green algal flora of Rajasthan. *Nova Hedw.* **23**: 481-492.

Gupte, S.Y. 1964. Myxophyceae from the Ajee river near Rajkot (Saurashtra). *J. Indian Bot. Soc.* **43**: 9-16.

Heinbokel, J.F., 1986. Occurrence of *Richelia intracellularis* (Cyanophyta) within the diatoms *Hemiaulus haukii* and *H. membranaceus* of Hawaii. *J. Phycol.* **22**: 399-403.

Hoffmann, J., 1989. Algae of terrestrial habitats. *Bot. Rev.* **55**: 77-105.

Holm-Hansen, O., 1968. Ecology, physiology and biochemistry of blue-green algae. *Ann. Rev. Microbiol.* **22**: 47-70.

Khoja, T.M. and B.A. Whitton, 1975. Heterotrophic growth of filamentous blue-green algae. *Br. Phycol. J.* **10**: 139-148.

Khurana,A.S. and G.S. Venkataraman, 1968. Algal and fungal flora of paddy soils. *Indian J. Microbiol.* **8**: 91-93.

Kumar, H.D., 1990. *Introductory Phycology*. Affiliated East-West Press Pvt.Ltd., New Delhi. 386 p.

Lee, R.W., 1989. *Phycology*. Cambridge University Press, Cambridge. 646 p.

Lewin, R.A. and P.T. Robinson, 1979. The greening of polar bears in zoos. *Nature***278**: 445-447.

Mague, T.H., 1977. Ecological aspects of dinitrogen fixation by blue-green algae. In : "A Treatise of Dinitrogen Fixation". Section IV. Hardy, R.W.F. and A.H. Gibson, (eds.)John Wiley and Sons, Inc. p. 85-140.

Marathe, K.V. and P.R. Chaudhari, 1975. An example of algae as pioneers in the lithosphere and their role in rock corrosion. *J. Ecol.* **63**: 65-70.

Misra, J.N., 1960. The ecology, distribution and seasonal succession of the littoral algae on the West coast of India. In : Proc. "Symposium on Algology". (Kachroo, P., ed.) I.C.A.R., New Delhi. p. 187-203.

Mitra, A.K., 1950. Two new algae from Indian soils. *Ann. Bot.* 14: 457-464.

Palmer, C.M., 1980. *Algae and Water Pollution.* Castle House Publications Ltd., England, 123 p.

Pandey, D.C., 1966. A study of the algae from paddy soils of Ballia and Ghazipur districts of Uttar Pradesh, India - 1. Culture and ecological conditions. *Nova Hedw.*9: 299-303.

Peterson, J.B., 1935 Studies on the biology and taxonomy of soil algae. *Dansk. Bot. Ark* 8: 1-180.

Prasad, B.N., and M.N. Srivastava, 1984. On a teratological *Scytonema* Ag. from Andaman Islands. *Phykos* **23**: 1-2.

Prasad, B.N., and P.N. Srivastava, 1965. Thermal algae from Himalayan hot springs. *Proc. Nat. Insti. Sci., India.* 31: 45-53.

Prasad, B.N., and Y. Singh, 1980. Algal hydrobiology in India. *Proc. Natl. Acad. Sci.* **50:** 271-300.

Rai,A.N., ed., 1990. *Handbook of Symbiotic Cyanobacteria.* CRC Press.

Raju, N.S., 1964. *Aulosira terrestre* sp. nov., occurring in the soils of Hyderabad (India). *Phykos* 3: 33-34.

Rose, E.T., 1934. Notes on the life history of *Aphanizomenon flos-aquae*. *Univ. Lowa Studies Nat. Hist.* **16**: 129-140.

Shapiro,J., 1973. Blue-green algae : why they become dominant? *Science* 179: 382-384.

Singh, A.L. and P.L. Singh, 1989. *Nitrogen-fixation in Indian Rice Fields.* Agro Botanical Publishers (India), Bikaner, 236 p.

Singh,R.N., 1942. *Wollea bharadwajae* sp. nov. and its autecology. *Ann. Bot.* 6: 593-606.

Singh, R.N., 1950. Reclamation of usar lands in India through blue-green algae. *Nature* 165: 325-326.

Singh, R.N., 1961. *Role of Blue-green Algae in Nitrogen Economy of Indian Agriculture.* I.C.A.R., New Delhi. 175 p.

Srinivasan, K., 1960. Distribution patterns of marine algae in Indian seas. In: Proc. "Symposium on Algology". (Kachroo,P.,ed.) I.C.A.R., New Delhi. p. 219-242.

Stanier, R.Y., M. Kunisawa, M. Mandel and G. Cohen-Bazire, 1971. Purification and properties of unicellular blue-green algae (Order Chroococcales). *Bact. Rev.* **35**: 171-205.

Stanier, R.Y., W.R. Sistrom, T.A. Hansen, B.A. Whitton, R.W. Castenholz, N. Pfennig, V.N. Gorlenko, E.N. Kondratieva, K.E. Eimhjellen, R.Whittenbury,R.L. Gherna and H.G. Truper, 1978. Proposal to place the nomenclature of the Cyanobacteria (blue-green algae) under the rules of the International code of Nomenclature of Bacteria. *Int. J. Syst. Bacteriol.* **28**: 335-336.

Thomas, J. and E.A. Gonzalves, 1965. Thermal algae of Western India.1. Algae of hot springs at Akloli and Ganeshpuri. *Hydrobiologia.* **25**: 330-340.

Thompson, R.H., 1972. Algae from the hair of the sloth *Bradypus. J. Phycol.* 8 (Suppl.): 8.

Trainor, F.R., 1985. Survival of algae in a desiccated soil: a 25 year study. *Phycologia* **24**: 79-82.

Tripathi, S.N. and E.R.S. Talpasayi, 1980. Sulphydryls and survival of subaerial algae. *Curr. Sci.* **49**: 31-32.

Vashishta, P.C., 1965. More Cyanophyceae from Hoshiarpur. III. *J. Bombay Nat. Hist. Soc.* **62**: 104-119.

Vashishta, P.C., 1966. Thermal Cyanophyceae of India. *Ph.D Thesis,* Panjab University, Chandigarh.

Venkataraman, G.S., S.K. Goyal, B.D. Kaushik and P.Roychoudhury, 1974. "Algae : Form and Function". Today & Tomorrow's Printers & Publishers, New Delhi. 562 p.

Varma, A.K., A.K. Mitra and G.S. Venkatarman, 1964. Role of *Aulosira fertilissima* in the nitrogen economy of the soils. *Phykos.* **3**: 29-32.

Vicencio, Z.T., 1977. Studies on the food habits of milkfish, *Chanos chanos* Forks. *Fish. Res. J. Phil.* **2**: 3-18.

Villareal, T.A., 1989. Division cycles in the nitrogen-fixing *Rhizosolenia* (Bacillariophyceae) - *Richelia* (Nostocaceae) symbiosis. *Brit. Phycol. J.* **24**: 357-365.

Walsby, A.E. 1972. Structure and function of gas vacuoles. *Bact. Rev.* **36**: 1-32.

Ind. Phyco. Rev. 1993.
2: 59-72.

HEAVY METAL BIOLOGY

S.P. SINGH
C.A.S. in Botany, Banares Hindu University,
Varanasi - 221 005, U.P.

Indian work on heavy metal toxicity to algae seems almost fragmentary and has been limited to a few centres engaged in phycological studies. As it stands today, the main and major contribution to Indian work has emerged from the Banares (now, Varanasi) School of Phycology. We (the author and co-workers) owe a great deal to late Prof. R.N. singh for guiding the early footsteps in phycology. It was in the later 70's that Mn toxicity and mutagenesis in *Chroococcus limneticus* and *Plectonema boryanum* was reported (Singh and Kashyap, 1978). In a follow up programme, Cd-toxicity and the possible mechanisms of development of tolerance/resistance have been worked out in the rice field cyanobacterium *Nostoc calcicola* (Singh and Pandey, 1981a, b, c, 1982; Singh *et al.*, 1982, 1985). The cyanobacterial cells challenged with low Cd levels (2.0 µg/ml), produced truly branched structures, quite atypical of the nostocacean pattern of differentiation. Such abnormalities were reversible following a few division cycles in a medium devoid of Cd. While populations exposed to Cd ceased utilization of nitrogen sources (NO_3^- and NH_4^+), the cells under Cd-stress showed increased phosphate uptake (Singh and Pandey, 1981 a). Synthetic complexans like EDTA (0.05 mM) lessened Cd-toxicity in contrast to citrate (5.0 mM); amino acid, glutamine (1.0 mM), cysteine (0.05 mM) or alkalinity were antagonistic to Cd, and arginine (1.0 mM), aspartic acid (1.0 mM) contributed towards extended survival against elevated levels of Cd. One of the Cd-resistant strains of *N.calcicola* showed cross-resistance to antibiotics like neomycin (1.0 µg/ml), chloramphenicol (3.0 µg/ml), but not to streptomycin and among the metals, to Zn (20.0 µg/ml) and Hg (1.0 µg/ml). On the basis of such findings, it has been suggested that Cd-resistance may also be determined by gene(s) controlling antibiotic resistance in cyanobacteria (Singh and Pandey, 1982). Cd ions also induced the loss of intracellular K^+ through the damaged cell membranes of *N.calcicola* (Singh *et al.*, 1985).

The subsequent investigations were devoted towards unicellular cyanobacterium *Anacystis nidulans* (Singh and Yadava, 1983, 1984, 1985, 1986, 1987). Cd ions competitively inhibited NO_3^- uptake, and this inhibition was further enhanced in the presence of Hg, due to synergistic interactions, and protected when Ca and Zn were present (Singh and Yadava, 1983). Experiments on similar lines revealed that the Cd-inhibition of NH_4^+ and PO^{3-}_4 uptake was of non-competitive type. The Cd + Zn or Cd + Ca combinations, in general, reduced Cd-inhibition of NH^+_4 and PO^{3-}_4 uptake in contrast to Hg (Singh and Yadava, 1984). The uptake of Cd in the cyanobacterium was concentration-dependent, inhibited by Ca and Zn, and enhanced by Hg, and regulated by redox potential, thiols, EDTA and citrate. There was more Cd uptake in dense *A.*

nidulans cultures, but the specific rate on per unit protein basis, was less (Singh and Yadava, 1985). The Cd-tolerant strains of *A. nidulans* showed reduced electrolyte efflux and Cd uptake compared to the parent strain (Singh and Yadava, 1986). The amelioration of Cd-toxicity was remarkable in the presence of extracellular metabolites of the cyanobacterium (Singh and Yadava, 1987).

The subsequent attention has been directed to explore the possible action site(s) of Hg, Cu, Cd or Zn in cyanobacterial photosynthesis (Singh, C.B. and Singh, 1987; Singh, DP. and Singh, 1987). *Nostoc calcicola* cells pre-exposed to Hg (1 h), showed extreme sensitivity of selected photochemical and biochemical reactions of photosynthesis. Hg in combination with either Cd or methylmercury, interacted synergistically in contrast to the antagonism with either Ni or Cu in inhibiting photosynthetic O_2-evolution and $^{14}CO_2$-incorporation. It is suggested that: (a) Hg inhibited cyanobacterial photosynthesis at the very initial stage, (b) photosynthetic O_2-evolution was more sensitive to Hg than $^{14}CO_2$-uptake, and (c) Hg ions predominantly attack the action sites of Mn at PS II and ATP generation steps of photophosphorylation (Singh, C.B. and Singh, 1987). Likewise, Cu, Cd and Zn were also inhibitory to O_2-supported Hill activity and photosynthetic O_2-evolution in the membrane preparations of *Anacystis nidulans* (Singh, D.P. and Singh, 1987). The reversal of Cd and Zn inhibition, and not of Cu by exogenously added catalase suggested that Cd and Zn were inhibitory to the degradation of H_2O_2. Ascorbic acid supported 27% of the Hill activity, which was insensitive to DCMU and the remaining activity was not only sensitive to DCMU but also to Cu ions. Therefore, it has been suggested that the action site of Cd and Zn is located between the electron donation sites of H_2O_2 and ascorbic acid, while that of Cu, is located beyond it. Further, the electron donation by reduced glutathione was insensitive to DCMU and Cu, indicating that the action site of Cu is prior to its electron donation site. Also, the phenanthroline-reversal of Cu inhibition of Hill activity suggested a tentative action site of Cu at the cytochrome reductase level.

The author's laboratory has also been engaged in looking for the action site(s) of metal cations in the cyanobacterial N_2- metabolism (Singh *et al.*, 1987). The Hg and Cu-induced inhibition of *in vivo* glutamine synthetase and nitrogenase activity has been attributed to be a consequence of heavy metal-induced depletion of cellular ATP, reductants and carbon skeletons. Interactions between Hg and Ni, Cu, Cd or CH_3Hg with reference to either GS or N_2ase activity revealed that Ni and Cu were antagonistic to Hg while Cd and CH_3Hg interacted synergistically.

Our attention has been towards Cu-toxicity and uptake in free and immobilized cyanobacterial cells (Singh and Verma, 1988; Singh, S.P. *et al.*, 1989; Verma and Singh, 1990). Cu uptake in free and immobilized cells of *N. calcicola* revealed that Cu uptake rate was invariably higher in the latter over the former. The total Cu uptake in immobilized cells was almost 2.5-times more than their free cell counterparts. Also, the immobilized cells showed a stronger positive correlation between Cu adsorption and uptake (Singh S.P. *et al.*, 1989). Cu uptake in free cells was typically biphasic, comprising rapid binding of the cations to the cell wall, followed by the subsequent metabolism-dependent intracellular uptake. The cellular Cu uptake was light - and ATP-dependent, and the addition of DCMU or exogenous ATP established that the energy to drive Cu transport was mainly coming from PS II reactions. Using various metabolic inhibitors, it has been suggested that -SH group(s), proton gradient across the cell membrane, and ATP hydrolysis were involved in the transmembrane movement of Cu in *N. calcicola* (Singh and Verma, 1988; Verma and Singh, 1990). That metal cations induced uncontrolled ion efflux in biomembranes, was certified from membranes of *N. calcicola* (Singh and Singh, 1990). Also, the electrolyte efflux was directly proportional to Hg concentrations and exposure time. A comparison of NO_3^- uptake and *in vivo* nitrate reductase activity revealed that the former was 5-times more sensitive to Hg. It has been suggested that : (i) a significant correlation exits among such parameters, (ii) Hg predominantly attacks the cyanobacterial cell membrane, (iii) inhibits NO_3^- utilization and (iv) the metal toxicity is regulated by presence of other heavy metal cations (Singh and Singh, 1990). Experiments covering Ni regulation of photoautotrophy in *Nostoc muscorum* have shown that Ni inhibited photoautotrophic growth, nitrogenase activity photosynthetic O_2-evolution, $^{14}CO_2$-uptake and respiratory O_2-uptake; $^{14}CO_2$-uptake being more sensitive to Ni compared to N_2ase, O_2-evolution or O_2-uptake. Lower concentrations (1.0-5.0 μM), on the other hand, stimulated N_2ase as well as ribulose bisphosphate corboxylase activity (Asthana *et al.*, 1990). It is proposed that Ni inhibition is mainly through the regulation of RuBPcase activity in *N. muscorum.*

Elements like Co, Ni, Hg, Zn and Mn have been widely used to assess their toxicity in cyanobacteria (see Kashyap and Johar, 1984; Singh, 1989; Singh and Rai, 1990). Co induced inhibition of general photoautotrophic growth of *Anabaena doliolum* and *Anacystis nidulans,* and such inhibitions were partly or in some case, completely reversed by sulphur-containing amino acids (Singh, 1985). Ni, not included in the list of microelements for cyanobacterial growth, stimulated growth of *A. doliolum* and *A. nidulans* at a concentration of 0.01 μM when urea was supplemented as the nitrogen source, however, high concentrations were inhibitory. Ni-toxicity could be overcome by the addition of citrate as the chelating agent. Ni (0.01 - 20.0. μM) in the uptake medium, did not affect urea uptake to a significant extent in *A. nidulans,* nevertheless, a slight stimulation was observed in *A. doliolum* in the concentration range of 0.01 - 0.1 μM (Singh and Rai, 1990). Effect of Cu on ammonium transport in *Nostoc calcicola* with reference to interaction with sulphhydryl agents has been investigated by Kashyap and his students (Kashyap and Johar, 1984). Cu (0.1 μM) stimulated NH_4^+ uptake by the high-affinity system but similar effects were not observed in case the equimolar concentrations of either Hg, Ni, Zn or Mn were used. Since the inhibition of NH_4^+ uptake by the sulphhydryl blocking agents NEM, (N-ethylmaleimide) or PCMB, (p-chloromercuribenzoate) was not reversed by Cu, it was suggested that the oxidation of available sulphhydryl residues in membrane proteins is an important factor in NH_4^+ translocation in *N. muscorum* (Kashyap and Johar, 1984). Further work on this aspect was carried out by S.L. Gupta, who is currently at B.S.I. Calcutta (see Gupta, 1986) and D.P. Singh, now at Bhopal University, Bhopal (see Singh D.P. *et al.*, 1989).

Cu-toxicity and transport in the unicellular cyanobacterium *Anacystis nidulans* with respect to varying pH levels has been studied by Singh (1985). The observations revealed that the saturating Cu concentration for Michaelis-Menten kinetics was 25.0 µM and no characteristic pH profile could be obtained for Cu-uptake in a pH range of 4-9; the peak uptake rate was at pH 4. The sensitivity of Cu uptake to protonophore FCCP (p-trifluoromethoxyphenyl hydrazone), suggested the involvement of a proton gradient. The metal uptake remained uninhibited in dark or in the presence of DCCD (N,N'-dicyclohexycarbodiimide), indicating that energy derived through ATP hydrolysis may not be directly involved in the process. Further studies suggested the genetic control of Cu uptake in *A. nidulans* as mutants differed among themselves as well as from the parent type in terms of 'low' and 'high' affinity uptake systems (Singh, 1985). Similar Cu effects were also observed in this cyanobacterium in respect of growth, photosynthetic pigments and macromolecules (Gupta, 1986).

Also, an analogous genetic control of Mn transport in *A. nidulans* has been reported subsequently (Singh, 1986). Mn uptake by wild type *A. nidulans* showed saturation at 27.0 µmol/l. The uptake values at different pH (4.0-9.0) exhibited a characteristic profile with an optimum at 6.0. Almost similar Mn uptake rate in dark and light suggested that the energy needed for transport of the cation may not be directly derived from light reactions. D.P. Singh is currently engaged in exploring the action sites of Ni, Hg and Cu on the photochemical activities of cyanobacteria at Bhopal (see Singh D.P. *et al.,* 1989). Hg, Ni and Cu individually inhibited growth, O_2-evolution and PS I activity (O_2-uptake) in the cyanobacterium *Cylindrospermum* in the sequence: Hg > Ni > Cu. Such cations also inhibited the variable fluorescence of chlorophyll *a* and DCPIP reduction, which could be reversed by the protonophore FCCP. This led him conclude that heavy metal-induced modulation of charge separation (H^+ ion movement) across cell membrane might account for an increase in the high energy state quenching of fluorescence and inhibition of Hill activity. The addition of electron donors (DPC, (1,5-diphenycarbazide), H_2O_2) could not release the inhibition of Hill activity by either Hg, Ni or Cu. In contrast, ascorbic acid ferricyanide supported cytochrome *c* reduction was greatly stimulated by the lethal doses of Hg or Cu, but not of Ni. Such cations were strong inhibitors of ascorbic acid + NADP reduction with the exception of electron transfer from PMS (N-methyl phenazine methosulphate) to NADP that was sensitive to Ni only. Such findings suggested that action sites of Hg and Cu were located prior to the electron donation site of PMS, possible after cytochrome reductase, where such cations might be accounting for the increased level of reduced cytochrome by preventing its simultaneous oxidation. The inhibition site of Ni, however, was locted beyond the site of PMS (Singh D.P. *et al.,* 1989).

Agrawal and Kumar (1975) reported the mode of action of Co in the physiology and biochemistry of *Anacystis nidulans* (Agrawal and Kumar, 1977). The subsequent studies included the effect of vanadium, tungsten and copper on *Nostoc linckia* (see Mishra and Kumar, 1984; Kumar *et al.,* 1985). Vanadium stimulated growth and nitrogenase activity of molybdate - limited *N. linckia* cells but not the nitrate reductase activity. Tungsten, on the other hand, inhibited all the three processes. Copper was inhibitory to growth, nitrogenase activity and H_2 production in the cyanobacterium. The overall data suggested that copper pollution could considerably alter nitrogen dynamics in freshwaters (Kumar *et al.*, 1985).

Various physiological characteristics were studied to understand the mechanism of Ni and Ag toxicity and these include, general photoautotrophic growth, survival, $^{14}CO_2$ fixation, heterocyst differentiation and nitrogenase activity. All such processes were either inhibited or reduced considerably accompanied by the increased cellular loss of K^+ and Na^+. Ascorbic acid, glutathione and L-methionine offered protection against Ag and Ni. However, the level of protection by cysteine was quite different, i.e., less than 27% for Ag. It has been suggestd that protection against Ag and Ni toxicity in *Nostoc muscorum* by selected amino acids lends further support to a self-detoxifying ability of cyanobacteria as these procaryotes also synthesize all essential amino acids (Rai and Raizada, 1985, 1986, 1987; Rai, 1989). Similar comparisons were also made with regard to Cr and Pb toxicity in *N. muscorum*. Approximately 50% of the cyanobacterial populations survived 68.0 μM Cr and 72.0 μM Pb. The sublethal concentration of Cr inhibited approximately 82% of $^{14}CO_2$ uptake and 64% of acetylene reduction in comparison to Pb that inhibited these processes by only 15% and 36% respectively. Characteristically, toxicity in such cases was found to be reversed by only Ca and not EDTA, thus suggesting that Ca plays an important role in ameliorating Cr and Pb toxicity to *N. muscorum* (Raizada and Rai, 1985). The same cyanobacterium could be protected against Cr and Pb by ascorbic acid, reduced glutathione and sulphur-containing amino acids. Appreciable stimulation of selected metabolic events by ascorbic acid and glutathione was also noticed in the untreated cyanobacterial cultures. Sulphur-containing amino acids (at 5 μg/ml), on the othe, hand, exerted little stimulatory effect on untreated cultures. The results speak in favour of sulphur-containing compounds and reducing agents that could protect cyanobacterial cells against Cr and Pb even in nature, possibly by providing a reducing power that prevents oxidation of enzymes (Rai and Raizada, 1988, 1989).

Toxicity tests involving Cr and Sn were also extended to *Anabaena doliolum* taking parameters like, NO_3^- uptake, NH_4^+ uptake, nitrate reductase, glutamine synthetase, $^{14}CO_2$-uptake and heterocyst differentiation and the regulation of toxicity by cations, like Ca, Mg, Mn, Ni, Co or Zn, pH, salinity, complexans, carbon sources and sulphur-containing amino acids (Dubey and Rai, 1987, 1989; Rai and Dubey, 1988, 1989a,b). Ca, Mg and Mn substantially antagonized Cr and Sn effects with reference to growth, NO_3^- and NH_4^+ uptake in the hierarchical sequence as : Ca> Mg> Mn, whereas the sequence could be altered as Mn> Mg> Ca for nitrate reductase and glutamine synthetase activities. A synergistic interaction was noted for all the parameters when Ni, Co, or Zn were used in combination as : Ni> Co> Zn (Rai and Dubey, 1989; Dubey and Rai, 1990a,b). Alkalinity lowered the toxicity of Cr and Sn to *A. doliolum* compared to acidic conditions. Salinity (NaCl at 20 mM), extracellular metabolites and organic acids reduced the metal toxicities. Among the various organic acids, humic acid was most effective, apparently because of its multiple metal binding sites (Dubey and Rai, 1987, 1989; Rai and Dubey, 1988, 1989b).

Further investigations on *Anabaena doliolum* and *Nostoc muscorum*, suggested that the loss of cellular electrolytes can be used as a first signal of heavy metal toxicity (Mallick and Rai, 1990; Rai *et al.*, 1990b). Heavy metal studies on *Chlorella vulgaris, A. doliolum* and *A. flos-aquae,* regulation by certain environmental factors and sequential addition of Fe, Ni and Cu in the test medium, were also carried out. These coupled with the morphometric and X-ray energy dispersive analyses suggested : (i) $^{14}CO_2$-incorporation is the most suitable variable to assess heavy metal toxicity, (ii) length of exposure and sequence of metal addition are crucial in monitoring of the metal toxicity, (iii) pH, salinity, extracellular metabolites and organic acids can mediate the toxicity in a laboratory microcosm, and (iv) such studies may be ideal for bioassay studies in algae (Mallick and Rai, 1989; Rai *et al.*, 1990a; Mallick *et al.*, 1990; Dubey and Rai, 1990).

Currently, the group is engaged in the use of structural and functional variables *in situ* through CEPEX enclosures for the assessment of heavy metal toxicity in the phytoplankton community of river Ganga at Varanasi (Rai *et al.*, 1990; Singh and Rai, 1990, 1991). The detailed study includes effect of selected heavy metals on autotrophic index, pigment diversity, ^{14}C uptake and *in situ* nitrogenase activity. Cd, Zn, Hg or Cr (8.0 μg/ml) resulted in maximum reduction in algal populations. The filamentous forms showed greater tolerance against Cd or Zn, compared to unicellular forms. With the exception of unicellular *Anorthoneis excentrica,* others were extremely sensitive to Hg or Cr. The decline in algal number depended on metal concentration and metal type with differences at generic and species level. Such observations revealed that (i) the variable $^{14}CO_2$ uptake could be adopted for assessment of heavy metal toxicity in an aquatic ecosystem, and (ii) heavy metals inhibit both structural and functional variables of phytoplankton in field microcosms. Therefore, this technique holds potential for use in biomonitoring of heavy metal toxicity in the field (Singh and Rai, 1990, 1991).

Further work is in progress to account for the precise mechanism of Hg, Zn and Cd action in photosynthetic and associated electron transport processes in *Nostoc linckia* (Husaini *et al.*, 1991). The recent investigations have compared the relative metal-sensitivity of PS I, PS II and ATP pool potential in green and blue-green algae.

Tripathy and Mohanty (1981) reported that the action of heavy metals on photosynthetic electron transport varied greatly from metal to metal. For example, nickel inactivated the chloroplast photosynthetic electron transport activity (Tripathy *et al.*, 1981), while cobalt without affecting the photosystem I, inhibited the activity of photosystem II (Tripathy *et al.*, 1983). The cations stimulated electron transport associated with PS II on one hand and , inhibited the electron flow linked with PS I the other in sphaeroplasts of unicellular cyanobacterium *Synechococcus cedrorum* (Wavare and Mohanty, 1983). In late 80s, Mohanty group has concentrated mainly on mercury-induced inhibition of PS II activity and changes in the emission of fluorescence from phycobilisomes and chlorophyll *a* in *Spirulina platensis* (Murthy *et al.*, 1989, 1990; Murthy and Mohanty, 1990). Addition of even low concentrations of $HgCl_2$ to intact cells of *S. platensis* caused an enhancement in the intensity of fluorescence emitted from phycocyanin at room temperature, and induced 'blue shifts' in the emission peak suggestive of changes in energy transfer within the phycobilisomes. Such ions also suppressed the whole-chain electron transport activity ($H_2O \rightarrow$ methylviologen) at fairly low concentrations than that required to inhibit Hill activity supported by p-benzoquinone. The extent of inhibition of Hill activity was much higher under high than under low intensity light. It has been inferred that Hg ions at low concentrations, affect the transfer of energy within phycobilisomes and at high, inhibit electron transport in the cyanobacterium (Murthy *et al.*, 1989). Quite recently, Murthy *et al.* (1990) observed three distinct change in the fluorescence yield of chlorophyll *a*, that depended on mercury load: (a) at low concentration (1.5 µM), $HgCl_2$ behaves in a manner similar to that of diuron in increasing the fluorescence intensity in weak, modulated light, which was attributable to the blockage of electron flow on the reducing side of PS II, (b) at slightly increased levels (3.0 µM), the quenching of the variable fluorescence of chlorophyll suggested a decrease in the extent of electron flow on the donor side of PS II; this effect becomes much more evident in presence of diuron, and (c) sufficiently high concentration (18.0 µM) resulted in the pronounced quenching of chlorophyll fluorescence that might have resulted through blockage of PS II on the donor side and the structural changes in antena pigments. Such multiple effects of Hg are also observed in intact cells of *Anacystis* and in sphaeroplasts theirof, although the latter required only lower Hg concentrations than the former. As the effect of Hg could not be reversed by EDTA, it has been suggested that the metal binds with the pigment-protein complexes of the cyanobacterium.

Azeez and Banerjee (1986) studied Cu, Ni and Cr-toxicity in terms of carbon assimilation, chlorophyll *a* content and several other parameters in genera like *Anacystis nidulans* and *Spirulina platensis*. The group has suggested that heavy metal accumulation by cyanobacteria is influenced by various factors of the micro-environment. The metal uptake events include : (a) passive mechanism, where the cell surface of the organism fuction as an ion exchange site for metal cations prevailing in the ambient medium, and (b) the active intracellular transport mechanism. The physico-chemical factors influencing metal accumulation are : (a) metal speciation, (b) the binding characteristics, and (c) the metabolism of the cyanobacterium. Complexans, both natural and artificial, influence significantly the metal accumulation/toxicity. However, no generalization can be made about the mechanism(s) involved. Many of the algae synthesize and release complexans in order to capture the metal micronutrients (Azeez *et al.*, 1987; Sharma *et al.*, 1987).

Gupta and Arora (1978) from Kanpur, observed several morphological as well as physiological aberrations in the cyanobacterium *Lyngbya nigra* exposed to deleterious concentrations of Cu. Algal assay technique has been successfully deployed to define trace elements limitation and heavy metal toxicity, by a group working at Dharwad (Bharati *et al.*, 1987). The co-existence of heavy metals affect the toxic action(s) of a particular heavy metal under consideration, e.g., Cd-toxicity to biological systems was considerably modified in the presence of other metal cations (Sharma *et al.*, 1985). Several factors are known to be involved in the mobility of heavy metal ions in terrestrial and aquatic ecosystems as well. Such investigations pertain to the mobility of selected heavy metals through Indian Red Soil (Khan *et al.*, 1982) or in aquatic systems, like river waters of Ganga at Varanasi (Mathur *et al.*, 1987).

Reddy and Venkateswaralu (1985) have made an assessment of the water quality and pollution load in the river Tungbhadra at Hyderabad, taking several physicochemical factors, heavy metal input and phycological characteristics. The river water receiving paper mill effluents, showed increase in the contents of bicarbamates, chlorides, phosphates, sulphates, total hardness and dissolved solids. The occurrence, structure, synthesis and function of metal binding proteins/peptides of algae have been reviewed by Reddy and Prasad (1990). Metal-binding proteins which are produced in response to various heavy metals in plants (including algae), are quite different from animal metallothionein in structure and synthesis. Such proteins isolated and characterized by the group, comprised mainly of three amino acids, i.e., cysteine, glutamic acid and glycine. The peptides have the general structure of (γ-glutamyl-cysteinyl)-glycine. These γ-glutamyl peptides have been named as phytochelatins, cadystins and γ-glutamyl peptides and are suggested to be involved in metal ion homeostasis and detoxification of heavy metals.

Acknowledgements

The author is grateful to Dr. C.B. Singh, Research Associate of CSIR (9/13 (307)/88/EMR-I, dt. 19.4.89) for his valuable and untiring efforts that have made the compilation work possible in the present form.

REFERENCES

Agrawal, M. and H.D. Kumar 1975. Response of *Chlorella* to mercury pollution. *Indian J. Ecol.* **2:** 94-98.

Agrawal, M. Jr. and H.D. Kumar 1977. Cobalt toxicity and its possibie mode of action in blue-green alga *Anacystis nidulans. Beitr. Biol. Pflanzen.* **53:** 157-167.

Asthana, R.K. P.K. Pandey and S.P. Singh 1990. Nickel regulation of photoautotrophy in a cyanobacterium. *Water Air Soil Pollut.* **52:** 263-276.

Azeez, P.A. and D.K. Banerjee 1986. Effect of copper and cadmium on carbon assimilation and uptake of metals by algae. *Toxicol. Environ. Chem.* **12:** 77-86.

Azeez, P.A. and D.K. Banerjee 1987. Influence of light on Chl *a* content of blue-green algae treated with heavy metals. *Bull. Environ. Contam. Toxicol.* **36:** 1062-1069.

Azeez, P.A., S. Panigrahi and R.M. Sharma 1987. Toxicity of combination of heavy metals to cyanobacteria, In: *International Phycotalk Symposium,* Banaras Hindu University, Varanasi, 18-21 December, 1987 p. 3.

Bharati, A., Arvind and R.P. Saxena 1987. Use of algal assay to define trace element limitation and heavy metal toxicity. *J.Environ. Biol.* **2:** 11-18.

Dubey, S.K. and L.C. Rai 1987. Effect of chromium and tin on survival, growth, carbon-fixation, heterocyst differentiation, nitrogenase, nitrate reductase and glutamine synthetase activities in *Anabaena doliolum. J. Plant Physiol.* **130:** 165-172.

Dubey, S.K. and L.C. Rai 1989. Toxicity of chromium and tin to *Anabaena doliolum:* Interaction with sulphur containing amino acids and thiols. *Biol. Metals* **2:** 55-60.

Dubey, S.K. and L.C. Rai 1990a. Toxicity of chromium and tin to *Anabaena doliolum:* Interaction with bivalent cations. *Biol. Metals* **3:** 8-13.

Dubey, S.K. and L.C. Rai 1990b. Heavy metal toxicity in a N_2-fixing cyanobacterium *Anabaena doliolum:* Regulation of toxicity by certain environmental factors. *Biomed. Environ. Sci.* **3:** 240-249.

Gupta, A.B. and A. Arora 1978. Morphology and physiology of *Lyngbya nigra* with reference to copper toxicity. *Physiol. Plantar.* **44:** 215-220.

Gupta, S.L. 1986. Copper uptake and inhibition of growth, photosynthetic pigments and macromolecules in the cyanobacterium *Anacystis nidulans. Photosynthetica* **20:** 447-453.

Husaini, Y., A.K. Singh and L.C. Rai 1991. Cadmium toxicity to photosynthesis and associated electron transport system of *Nostoc linckia. Bull. Environ. Contam. Toxicol.* **46:** 146-150.

Kashyap, A.K. and G. Johar 1984. Ammonium transport in nitrogen fixing cyanobacterium *Nostoc muscorum*: interaction with copper and sulfhydryl agents. *J.Gen. Appl. Microbiol.* **30:** 267-275.

Khan, S.; D. Nandan and N.N. Khan 1982. The mobility of some heavy metals through Indian Red Soil. *Environ. Pollut.* 4: 119-126.

Kumar, D.; M. Jha and H.D. Kumar 1985. Copper toxicity in the freshwater cyanobacterium *Nostoc linckia. J. Gen. Appl. Microbiol.* **31:** 165-169.

Mallick, N. and L.C. Rai 1990. Effect of heavy metals on the biology of nitrogen fixing cyanobacterium *Anabaena doliolum. Toxicity Assessment* **5:** 207-220.

Mallick, N.; A.K. Singh and L.C. Rai 1990. Impact of bimetallic combination of Cu, Fe and Ni on growth rate, uptake of nitrate and ammonia, $^{14}CO_2$-fixation, nitrate reductase and urease activity of *Chlorella vulgaris*. *Biol. Metals* **2:** 223-228.

Mathur, A.; Y.C. Sharma; D.C. Rupainwar; B.C. Murthy and S. Chandra 1987. A study of river Ganga at Varanasi with special emphasis on heavy metal pollution. *Pollut. Res.* **6:** 37-44.

Mishra, D.K. and H.D. Kumar 1984. Effects of vanadium and tungsten on the nitrogen-fixing cyanobacterium *Nostoc linckia*. *Biol. Plantarum* **26:** 448-454.

Murthy, S.D.S; N.G. Bukhov and P. Mohanty 1990. Mercury-induced alterations of chlorophyll *a* fluorescence kinetics in cyanobacteria. *J. Photochem. Photobiol.* **6:** 373-380.

Murthy, S.D.S. and P. Mohanty 1990. Effect of sub-lethal concentrations of mercury on photochemical functions of photosystem II in cyanobacterium *Spirulina platensis*. In: *National Symposium on Cyanobacterial Nitrogen Fixation*. New Delhi, 29-31 January p. 36.

Murthy, S.D.S.; S.C. Sabat and P. Mohanty 1989. Mercury-induced inhibition of photosystem II activity and changes in the emission of fluorescence from phycobilisomes in intact cells of the cyanobacterium *Spirulina platensis*. *Plant Cell Physiol.* **30:** 1153-1157.

Rai, L.C. 1989. Silver toxicity to a nitrogen fixing cyanobacterium: Interaction with Ni, Cr and Pb. *Biol. Metals* **2:** 122-128.

Rai, L.C. and S.K. Dubey 1988. Chromium toxicity to a cyanobacterium: Possible role of carbon sources in toxicity amelioration. *Microbios* **55:** 193-203.

Rai, L.C. and S.K. Dubey 1989a. Impact of chromium and tin on nitrogen-fixing cyanobacterium *Anabaena doliolum*: Interaction with bivalent cations. *Ecotoxicol. Environ. Safety* **17:** 94-104.

Rai, L.C. and S.K. Dubey 1989b. Protective effect of certain natural and synthetic complexans on the toxicity of chromium and tin to a nitrogen fixing cyanobacterium *Anabaena doliolum*. *Biomed. Environ. Sci.* **2:** 271-281.

Rai, L.C.; T.E. Jensen and J.W. Rachlin 1990a. A morphometric and X-ray energy dispersive analysis approach to monitoring pH altered Cd toxicity in *Anabaena flos-aquae*. *Arch. Environ. Contam. Toxicol.* **19:** 479-487.

Rai, L.C. and M. Raizada 1985. Effect of nickel and silver ions on survival, growth, carbon fixation and nitrogenase activity in *Nostoc muscorum:* Regulation of toxicity by EDTA and calcium. *J. Gen. Appl. Microbiol.* **31:** 329-337.

Rai, L.C. and M. Raizada 1986. Nickel induced stimulation of growth, heterocyst differentiation, $^{14}CO_2$ uptake and nitrogenase activity in *Nostoc muscorum. New Phytol.* **104:** 111-114.

Rai, L.C. and M. Raizada 1987. Toxicity of nickel and silver to *Nostoc muscorum:* Interaction with ascorbic acid, glutathione and sulphur-containing amino acids. *Ecotoxicol. Environ. Safety* **14:** 12-21.

Rai, L.C. and M. Raizada 1988. Impact of chromium and lead on *Nostoc muscorum*: Regulation of toxicity by ascorbic acid, glutathione and sulphur-containing amino acids. *Ecotoxicol. Environ. Safety* **15:** 195-205.

Rai, L.C. and M. Raizada 1989. Effect of bimetallic combinations of Ni, Cr and Pb on growth, uptake of nitrate and ammonia, $^{14}CO_2$-fixation and nitogenase activity of *Nostoc muscorum. Ecotoxicol. Environ. Safety* **17:** 75-85.

Rai, L.C.; M. Raizada; N. Mallick; Y. Husaini; A.K. Singh and S.K. Dubey 1990b. Effect of four heavy metals on the biology of *Nostoc muscorum. Biol. Metals* **2:** 229-234.

Rai, L.C.; A.K. Singh and N. Mallick 1990c. Employment of CEPEX enclosures for monitoring toxicity of Hg and Zn on *in situ* structural and functional characteristics of algal communities of river Ganga in Varanasi, India. *Ecotoxicol. Environ. Safety* **220:** 211-221.

Raizada, M. and L.C. Rai 1985. Metal induced inhibition of growth, heterocyst differentiation, carbon fixation and nitrogenase activity of *Nostoc muscorum:* Interaction with EDTA and calcium. *Microbios Lett.* **30:** 153-161.

Rana, B.C. and H.D. Kumar 1974. The toxicity of zinc to *Chlorella vulgaris* and *Plectonema boryanum* and its protection by phosphate. *Phykos* **13:** 60-66.

Reddy, G.N. and M.N. Prasad 1990. Heavy metal binding proteins/peptides: Occurrence, structure, syntheses and functions - A review. *Environ. Exp. Bot.* **30:** 251-264.

Reddy, P.M. and V. Venkateswarlu 1985. The assessment of water quality and pollution in the river Tungabhadra (A.P.), India. In: *All India Symp. on Biol. of Algae.* University of Poona, 8-10 Feb. 1985, p. 10.

Sharma, A.; A. Mukherjee and G. Talukdar 1985. Modification of cadmium toxicity in biological systems by other metals. *Curr. Sci.* **54:** 539-549.

Sharma. R.M.; P.A. Azeez and S. Panigrahi 1987. Role of complexans in metal accumulation and toxicity to phytoplankton: A review. In: *International Phycotalk Symposium*. Banaras Hindu University, Vananasi, 18-21 December 1987, p. 29.

Singh, A.K. and L.C. Rai 1990. Use of *in situ* structural and functional variables of phytoplankton of the river Ganga for assessment of heavy metal toxicity. *Biomed. Environ. Sci.* 3:397-405.

Singh, A.K. and L.C. Rai 1991. Cr and Hg toxicity assessment *in situ* using the structural and functional characteristics of algal communities. *Environmental Toxicology & Water Quality: An International Journal* **6**: 97-107.

Singh, C.B. and S.P. Singh 1987. Effect of mercury on photosynthesis in *Nostoc calcicola* : Role of ATP and interacting heavy metal ions. *J. Plant Physiol.* **129**: 49-58.

Singh, C.B. and S.P. Singh 1990. Hg^{2+}-induced leakage of electrolytes and inhibition of NO_3^- utilization in *Nostoc calcicola*: Role of interacting cations. *Biol. Metals* **3**: 208-212.

Singh, C.B.; S.K. Verma and S.P. Singh 1987. Impact of heavy metals on glutamine synthetase and nitrogenase activity in *Nostoc calcicola. J. Gen. Appl. Microbiol.* **33**: 87-91.

Singh, D.P. 1985. Cu^{2+} transport in the unicellular cyanobacterium *Anacystis nidulans. J. Gen. Appl. Microbiol.* **31**: 277-284.

Singh, D.P. 1986. Mn^{2+} transport in the unicellular cyanobacterium *Anacystis nidulans. J. Basic Microbiol.* **26**: 161-168.

Singh, D.P.; P. Khare and P.S. Bisen 1989. Effect of Ni^{2+}, Hg^{2+} and Cu^{2+} on growth, oxygen evolution and photosynthetic electron transport in *Cylindrospermum* PU 942. *J. Plant Physiol.* **134**: 406-412.

Singh, D.P. and S.P. Singh 1987. Action of heavy metals on Hill activity and O_2-evolution in *Anacystis nidulans. Plant Physiol.* **83**: 12-14.

Singh, S. 1989. Cobalt induced inhibition of growth in cyanobacteria *Anabaena doliolum* and *Anacystis nidulans* : Interaction with sulphur containing amino acids. *Indian J. Exp. Biol.* **27**: 1092-1093.

Singh, S. and A.K. Rai 1990. Nickel-dependent growth and urea uptake in the cyanobacteria *Anabaena doliolum* and *Anacystis nidulans. Indian J. Exp. Biol.* **28**: 80-82.

Singh, S.P. and A.K. Kashyap 1978. Manganese toxicity and mutagensis in two blue-green algae. *Environ. Exptl. Bot.* 18: 47-53.

Singh, S.P. and A.K. Pandey 1981a. Effect of cadmium ion on the differentiation and micronutrient utilization in *Nostoc calcicola. Ind. J. Microbiol.* **21:** 119-125.

Singh, S.P. and A.K. Pandey 1981b. Cadmium toxicity in a cyanobacterium: Effect of modifying factors. *Environ. Exp. Bot.,* **21:** 257-265.

Singh, S.P. and A.K. Pandey 1981c. Heavy metal interaction studies with a blue-green alga. *Acta Bot. Indica* **9:** 290-296.

Singh, S.P. and A.K. Pandey 1982. Cadmium mediated resistance to metals and antibiotics in a cyanobacterium. *Molec. Gen. Genet.* **187:** 240-243.

Singh, S.P.; A.K. Pandey and R.P. Singh 1982. Nitrogen fixation in the rice field cyanobacterium *Nostoc calcicola* exposed to cadmium and bisulphite. In: *Proc. Nat. Symp. on Biological Nitrogen Fixation,* DAE, Govt. of India, IARI, New Delhi, 1982, pp. 162-169.

Singh, S.P.; A.K. Pandey and R.P. Singh 1985. Efflux of cell potassium in a cyanobacterium exposed to cadmium and bisulphite. *Plant and Nature* **3:** 1-6.

Singh, S.P. and S.K. Verma 1988. Heavy metal uptake in the cyanobacterium *Nostoc calcicola. J. Ind. Bot. Soc.* **67:** 74-77.

Singh, S.P.; S.K. Verma; R.K. Singh and P.K. Pandey 1989. Copper uptake by free and immobilized cyanobacterium. *FEMS Microbiol. Lett.* **60:** 193-196.

Singh, S.P. and V. Yadava 1983. Cadmium induced inhibition of nitrate uptake in *Anacystis nidulans:* Interaction with other divalent cations. *J. Gen. Appll. Microbiol.* **29:** 297-304.

Singh, S.P. and V. Yadava 1984. Cadmium induced inhibition of ammonium and phosphate uptake in *Anacystis nidulans:* Interaction with other divalent cations. *J. Gen. Appl. Microbiol.* **30:** 79-86.

Singh, S.P. and V. Yadava 1985. Cadmium uptake in *Anacystis nidulans:* Effect of modifying factors. *J. Gen. Appl. Microbiol.* **31:** 39-48.

Singh, S.P. and V. Yadava 1986. Cadmium tolerance in the cyanobacterium *Anacystis nidulans. Biol. Zentralbl.* **105:** 539-5422.

Singh, S.P. and V. Yadava 1987. Factors affecting cadmium toxicity in *Anacystis nidulans. J. Environ. Biol.* **8:** 245-256.

Tripathy, B.C.; B. Bhatia and P. Mohanty 1981. Inactivation of chloroplast photosynthetic electron transport activity by Ni^{2+}. *Biochim. Biophys. Acta* **638:** 217-224.

Tripathy, B.C., B. Bhatia and P. Mohanty 1983. Cobalt ions inhibit electron transport activity of photosystem II without affecting photosystem I. *Biochim. Biophys. Acta* **722:** 88-93.

Tripathy, B.C. and P. Mohanty 1981. Stabilization of glutaraldehyde fixation of chloroplast membrane structure and function against heavy metal ion induced damage. *Plant Sci. Letter* **22:** 253-261.

Verma, S.K. and S.P. Singh 1990. Factors regulating copper uptake in a cyanobacterium. *Current Microbiol.* **21:** 33-37.

Wavare, R.A. and P. Mohanty 1983. Cation stimulate electron transport associated with photosystem II and inhibited electron flow linked photosystem I in sphaeroplast of the cyanobacterium *Synechococcus cedrorum. Photochem. Photobiol.* **6:** 189-200.

Ind. Phyco. Rev. 1993.
2: 73-82.

ALGAL BIOFERTILIZER TECHNOLOGY FOR RICE CROP

S. KANNAIYAN
Department of Agricultural Microbiology,
Tamil Nadu Agricultural University,
Coimbatore - 641003, Tamil Nadu.

INTRODUCTION

Nitrogen is usually the limiting factor to produce high rice yields. The green revolution in rice production is based on fertilizer responsive rice varieties. In Asia one of the constraints to high yields is the limited availability and high prices of nitrogen fertilizers. In 1978 fertilizer use in tropical Asian countries averaged 30 - 55 kg NPK/ha arable land (Palacpac, 1982). The idea of utilizing algal biofertilizer as an alternative or supplementary source of nitrogen for rice is a potential biological system under low cost rice production technology. The resulting nitrogen fertility by algal application has permitted moderate but constant productivity in fields where no N fertilizer is applied.

Biological nitrogen fixation requires energy generally obtained by the catabolism of photosynthetically fixed carbon, i.e. photosynthate. Among the N_2 - fixing microorganisms, only blue green algae are able to generate their own photosynthate from CO_2 and water. The agronomic potential of BGA was recognized in 1939 by De who attributed the natural fertility of tropical paddy fields to N_2 fixing BGA. Since, many field trials have been conducted to increase rice yield by algal inoculation or by cultural practices favouring the growth of indigenous BGA.

Higher nitrogen fixation in wetland condition has long been considered the reason for the higher nitrogen maintenance levels in wetland soils than dry land soils. The nitrogen balance of long-term fertility experiments gives a quantitative answer to the question of nitrogen gains in wetland condition (Watanabe and Roger, 1984). In submerged rice soil ecosystem, biological nitrogen fixation is essentially an algal process that contributes 25-30 kg ha^{-1} per cropping season (Venkataraman, 1972). The present day global interest in biological nitrogen fixation is a direct consequence of this necessity to provide some economic assistance to small and marginal farmers and to introduce a judicious combination of linear and cyclic fertilization of soils.

The beneficial effect of algal inoculation on the grain yield of rice has been well documented at a number of locations (Venkataraman, 1972; Jagannathan and Kannaiyan, 1977 and Kannaiyan, 1979). Algal inoculation to rice crop was reported to be effective in different agroclimatic conditions and soil types (Goyal and Venkataraman, 1971 and Kannaiyan, 1983). Blue green algal inoculation with composite cultures was reported to be more effective than single culture inoculation (Kannaiyan, 1978). A technology for mass-scale cultivation of composite culture of blue-green algae under rice field condition was developed (Kannaiyan, 1978) and BGA inoculation survive for more than 15 months in soil based composite culture (Kannaiyan *et al.*, 1980). At many sites where algal inoculation was used for three to four consecutive cropping seasons, the inoculated algae established well and the effect persisted over subsequent crops (Venkataraman, 1979).

The impact of blue-green algal application was lucidly reviewed by Venkataraman (1972). Biological N_2 fixation in flooded soil systems was reviewed by Buresh *et al.* (1980). Processes and ecology of N_2 fixation in rice soils were reviewed by Watanabe (1978) and Watanabe and Brotonegoro (1981). Technologies for utilizing nitrogen fixing organisms in low land rice were analysed by Roger and Watanabe (1985) in a review which emphasizes their potentialities, their current usage and the limiting factors. The beneficial role of blue green algal inoculation in rice soils of Tamil Nadu was recently documented by Kannaiyan (1985).

NITROGEN FIXING BLUE-GREEN ALGAE

The species of blue-green algae which are known to fix atmospheric nitrogen are classified into three groups (i) Heterocystous, aerobic forms, (ii) Aerobic, Unicellular forms and (iii) Non-heterocystous, filamentous, microaerophillic forms. The blue-green algal cultures composite inoculum consisting of *Nostoc, Anabaena, Tolypothrix, Plectonema, Aphanothece, Oscillatoria, Cylindrospermum, Aulosira* and *Scytonema* have been used for inoculation for rice. The algal cultures are able to survive in soil-based inoculum for more than 15 months (Kannaiyan, 1981).

APPEARANCE OF BGA IN THE CULTIVATION CYCLE OF RICE

Algal populations appear to be highly susceptible to environmental changes and exhibit rapid qualitative and quantitative variations along the cultivation cycle. Development of dense algal blooms just after transplanting due to fertilizers or ploughing or both and a high light availability have been reported. *Anabaena* was the first species to develop two weeks after transplanting (Roger and Kulasooriya, 1980). N_2 fixing forms, *viz. Scytonema, Aulosira, Nostoc, Aphanothece, Oscillatoria, Plectonema* were dominant during tillering phase of rice crop. During the cultivation cycle a decrease in light intensity and N level related to rice growth and an increase in pH value favoured BGA growth. *Gloeotrichia* and *Rivularia* are reported to occur during latter part of the cultivation cycle.

EPIPHYTIC BGA

Epiphytic habit of BGA have been observed in wetland rice (Roger *et al.* 1979), deep water (IRRI, 1977) and on weeds growing in rice fields (Kulasooriya *et al.* 1980). In the wetland rice field ecosystem, BGA epiphytic on rice and weeds make a limited contribution to the nitrogen input but play an important role in providing inoculum for the regeneration of algal blooms that are periodically affected by adverse conditions. In deep water rice, which offers a much greater biomass for colonization, the nitrogen contributed by the epiphytic BGA is agronomically significant. BGA were found to grow preferentially on sub-merged decaying tissues. An endophytic growth inside the leaf sheath, was also reported in deep water rice.

ALGAL INOCULATION TO RICE CROP

Application of dried blue-green algal flakes at the rate of 10 kg per hectare is recommended to the farmers. A composite starter culture consisting of more than one species or genera is more preferable. Further, it is always preferable to have the local flora of blue-green algae, as they acclamatise to the local surroundings much more easily than the introduced ones. Fast growing species which can tolerate certain amount of extremities are more suitable to slow growing and highly susceptible ones. The algal species should have a fair amount of tolerance to agricultural chemicals such as pesticides and weedicides. The algal inoculant is to be applied 10 days after transplantation of rice crop. Continuous application of blue-green algae to the rice fields for at least 3 or 4 consecutive seasons sustains the high crop yield at reduced levels of commercial nitrogen fertilizer input in the ensuing cropping seasons. The combined inoculation of algae and *Azolla* has performed better than individual inoculation (Kannaiyan *et al.* 1982).

RESIDUAL EFFECT OF BLUE-GREEN ALGAL INOCULATION

Blue-green algae are living bio-systems and once they establish their biological activity is a continuous process. Inoculation of blue-green algae for 3 or 4 cropping seasons continuously results in an appreciable population build up and consequently their beneficial effect can be seen in the subsequent years without any further inoculation, unless some unfavourable soil ecological conditions develop. The results of several fields trials conducted in Tamil Nadu have clearly demonstrated the positive residual effect of algal inoculation on rice yield (Kannaiyan, 1979). Field experiments carried out at the International Rice Research Institute, Philippines for 23 successive rice crops for 12 years without addition of nitrogenous fertilizers recorded no decline in rice yield, which was 3 to 4 tons per hectare per crop.

ALGAL INOCULATION ON SOIL PROPERTIES

Blue-green algae, besides carrying out the process of both photosynthesis and nitrogen fixation, add an appreciable quantity of bulk organic matter to the soil in the form of algal biomass, which is likely to influence the physical and chemical properties of the soil. In Indian paddy fields, the fresh weight of algal biomass produced is estimated to range from 200 to 1600 kg per hectare. Though the physical properties of the soil is not affected to any great extent, soil aggregation is found to be improved as a result of algal growth. The nitrogen content of the soil is also reported to be increased. A significant reduction in the contents of oxidisable matter, total sulphides and ferrous iron in the Kerala soils, as a result of successive algal inoculation has been reported. This is of particular local importance, since iron and sulphide toxicity is a common phenomenon in Kerala. This reduction is probably due to an increased oxygen tension in the medium resulting from algal photosynthesis.

SELECTION OF SUITABLE BGA STRAINS

Even when N_2 fixing indigenous strains are present in the soils, their efficiency may be poor and the introduction of new strains can be beneficial. There are large differences in the amount of N fixed by various genera and sometimes by the same species from different localities. The influence of genetic constitution might play a vital part in determining the capacity to fix nitrogen and the large variations in doubling time have been reported in different species. This may probably be related to strain abilities rather than to cultural conditions. Strain selected for use in the field should be fast growing and capable of fixing N_2 under aerobic, microaerobic and anaerobic conditions. They should also be able to grow photoautotrophically and chemoautotrophically and store endogenous carbohydrate reserves. They should evolve little H_2 and liberate nitrogen excess of their requirements of optimum growth. The ways in which glutamine synthetase of Blue-green algae is regulated could be of importance in strain selection. It is possible that the antagonistic effect of other organisms may affect the survival of a strain and the successful introduction of an effective strain may depend on its ability to survive and compete with the native flora for establishment, growth and effective N_2 fixation. It is also clear that the climate and physiological properties of the environment and resistance to pesticides will influence the efficieny of the inoculated BGA. These parameters have to be taken into account when selecting strains.

MODE OF NITROGEN AVAILABILITY FOR RICE

It is now well established that N_2-fixation by BGA plays a vital role in the buildup and maintenance of soil fertility but it is equally important to understand how much, when and in what ways the fixed N is made available to rice plants. Nitrogen fixed by the algae are released either through exudation or through microbial decomposition after the cells die (Roger and Kulasooriya, 1980). Laboratory experiments have frequently shown that BGA liberate large portions of their assimilated nitrogenous substances (Subrahmanian and Sahay, 1964). Release of nutrients through microbial decomposition after the death of the algae appears to be the principal means by which N is made available to the rice crop. Some algae are decomposed in 2-3 days, others with stand microbial digestion for more than 4 weeks. Laboratory experiments demonstrated that more than half of the nutrients contained in an algal biomass can be regenerated in less than 1 month with the aid of microbial degradation. A strain of *Bacillus subtilis* was found to decompose several N_2-fixing BGA very rapidly, about 40% of algal nitrogen being converted to ammonia within 10 days (Watanabe and Kiyohara, 1960). Grazers through their digestive tracts also make algal nitrogen available to rice. In paddy fields the death of algal biomass is most frequently associated with soil desiccation at the end of the cultivation cycle and algal growth has frequently resulted in a gradual buildup of soil fertility.

PRODUCTION OF GROWTH PROMOTING SUBSTANCES BY BGA

Besides increasing nitrogen fertility, BGA have been shown to benefit rice plants by producing growth promoting substances. More direct evidence for hormonal effects has come primarily from treatments of rice seedlings with algal cultures or their extracts (Roger and Kulasooriya, 1980). Pre-soaking of rice seeds in BGA cultures or extracts has decreased losses from sulfate reducing processes and this has been attributed to the enhancement of germination and a faster seedling growth due to algal exudates. Water soluble products from *Calothrix* sp., *Anabaena* sp. and *Stratonostoc* sp. had a rhizogenous effect and stimulated plant organs (Koptiyeva and Tantsiurenko, 1971). The probable nature of these substances has been linked to that of a gibberellin (Gupta and Shukla, 1972). It has been reported that the algal exudates contains substances like Vitamin B_{12}, gibberellic acid, Indole acetic acid and amino acids (Okuda and Yamaguchi, 1955). The production by BGA of substances that have a growth promoting effect on rice plants is well established but their mode of action is still unclear.

INFLUENCE OF PESTICIDES ON BGA

Under field conditions the toxicity of any pesticides depends on the initial microbial population, the nutrient status and the mode of application of pesticides. Pentachlorophenol incorporated in soil with lime stimulated N_2 fixing BGA, but if surface applied even at low levels, it was depressive with a long residual effect (Ishizawa and Matsuguchi, 1966). For non-persistent pesticides, the rate of degradation and the toxicity of products are important in considering the possible effects on algae. Resistance to pesticides varies widely with strain. *Cylindrospermum* sp. was reported to be less resistant to insecticides than *Aulosira fertilissima* and *Plectonema boryanum* (Singh, 1972). BGA can tolerate fungicide concentrations of 100 to 1000 ppm depending on the nature of both pesticides and strains tested (Gangawane and Saler, 1979). *Cylindrospermum* was reported to tolerate up to a maximum concentration of 800 ppm of 2, 4 D (Singh, 1972). *Aphanothece pallida* was found to tolerate even up to 1000 ppm of carbofuran and Phorate granules (Kannaiyan, 1980). The resistance to high levels of pesticides seems to be more characteristic of BGA than of the eukaryotic algae as indicated by a selective action of certain pesticides on mixed algal flora. Pentachlorophenol at 100 ppm was reported to be suppressed green algae without any deleterious effect on BGA. Gamma - BHC at a low level was reported to suppress diatoms but not the N_2 fixing BGA. Among the different groups of pesticides, insecticides have been reported to have a stimulatory affect on algal growth. At low concentrations a direct stimulatory affect of herbicides on growth and N_2 - fixation by *Tolypothrix tenuis* and *Calothrix brevissima* was reported. Low doses of insecticides at less than 10 ppm was reported to stimulate *Aulosira fertilissima* growth and Difolatan at 300 ppm also found to stimulate *A. fertilissima.*

RECLAMATION OF SALINE AND ALKALINE SOILS BY BLUE-GREEN ALGAE

The 'Usar' soils as it is locally termed in Uttar Pradesh representing the two groups of soils, *viz.,* saline and alkaline soils, are characterised by impermeability, extreme hardness and occasional presence of undesirable salts on the soil surface, which adversely affect plant growth. The apparent growth of algae in the 'Usar' soils is confined to the rainy season when the soil is moist. The first alga to appear is *Microcoleus chthonoplaste* in sparse patches over the moist efflorescence of sodium salts. Its value as a pioneer, therefore, lies in its soil-binding quality. The species, other than this associated in 'Usar' soils are *Microcoleus vaginatus, Scytonema ocellatum, S. javanicum, S. hofmanni, Porphyrosiphon notarisii, Campylonema lahorense, Cylindrospermum licheniforme, C. muscicola, Nostoc muscorum, N. liackia, N. commune, Anabaena ambigua, A. fertilissima, Wollea bharadwajae, Aulosira fertilissima.*

The algal biomass when incorporated in the soil conserves organic matter, nitrogen, phosphorus, moisture and subsequently converts the sodium clay into calcium type. It may be possible through the exchange complex of soils, as suggested in the conventional chemical amelioration through application of gypsum. Blue-green algae are often associated with the formation of calcareous deposits and many fresh water forms are capable of precipitating appreciable quantities of lime. They are also equally capable of solubilising such substrates. Though the entire process is not well understood, it is suggested that it may be due to secretion of oxalic acid. This characteristic on the part of many of the blue-green algae may be highly useful in reclaiming saline and alkaline soils.

MULTIPLICATION OF BLUE-GREEN ALGAE

Small scale multiplication in trays :

Shallow metal trays or cement tubs of size 9' x 6' x 3/4' can be used for algal multiplication on a small scale. A quantity of 5 kg of powdered field soil or tank silt is spread in the trays or tubs and water is added to a depth of 3 inches. Super phosphate at 150 g and lime at 30 g is added to each of the trays. The whole contents is mixed well and allowed to settle. A starter culture in the form of dry flakes, consisting of 6 to 10 different species of blue-green algae in a soil base is taken, powered and 250 g broadcasted over the water. Carbofuran granules at 25 g per tray is also applied to control soil insects. In about 7 to 10 days the algae develop into a thick mat in the tank. After 15 days of inoculation the water is allowed to dry up completely. The algal flakes are then collected and stored dry. From each tray a quantity of 5 to 7 kg of algal flakes is obtained.

Small scale multiplication in polythene lined pits :

The place selected for multiplication of blue-green algae must be in a open place and exposed to direct sunlight, since algae require sunlight for photosynthesis and nitrogen fixation. Pit of size 2 m x 1 m are marked out and dug to a depth of 20 cm, so that the sides slope inside. The pits are lined with polythene sheets to avoid percolation of water into the soil.

Five kg of field soil is added to each pit and filled with water to a height of 6 to 10 cm. Ferrous sulphate - 20 g, super phosphate - 100 g, potassium chloride - 25 g and slaked lime - 25 g are added to the pit, the whole thing mixed well and the contents allowed to settle for 8 to 12 hours. Carbofuran (Furadan) at 25 g was applied to each pit in order to eliminate soil insects. A starter culture of Blue-green algae at 250 g per pit is applied. Algal growth is seen from the fifth day after inoculation and in about 10 to 15 days the algae form a mat which floats on the surface of the water. In case the rate of evaporation of water from the pit is high, water is added to maintain the height of water at 6 to 10 cm. After 15 days of inoculation, the pit is allowed to dry up in sunlight and the dried algal flakes collected and stored. A yield of 4 to 7 kg is obtained from each pit.

Mass scale multiplication under field conditions :

The field selected for mass multiplication of blue-green algae is prepared well and levelled uniformly. The field is divided into small plots of size 10 m x 2 m (1/2 cent), by providing suitable bunds and irrigation channels. Water is allowed into a height of 10 cm. Super phosphate - 1500 g and lime - 150 g are added to each plot and inoculated with 1500 g of starter culture of blue-green algae. Carbofuran (Furadan) - 150 g is also added to prevent the activities of soil insects. Water is let in at periodical intervals, so that the height of water level is always maintained at 10 cm. The growth of algae is rapid and in about 10 days, a thick mat is formed which floats on the surface of the water. After 15 days of inoculation, the plots are allowed to dry up in the sun and the algal flakes collected and stored. From each harvest, 15 to 20 kg of dry algal flakes are obtained from each bed.

CONCLUSION

The flooded - soil - rice ecosystem is favourable for biological nitrogen fixation. Blue-green algae are ideal biological system under tropical situation for better nitrogen fixation and conservation in rice soil. Blue-green algae are renewable biofertilizer and they represent a self-supporting biological agents because they can photosynthetically provide energy for nitrogen fixation. Rice cropping system provides favourable aquatic environment for the growth, multiplication and nitrogen fixation by utilizing the solar energy. Thus, these potential group of nitrogen fixing algae may be better exploited for harnessing atmospheric nitrogen so as to increase the productivity of rice.

ACKNOWLEDGEMENT

The author is thankful to the Department of Biotechnology, Ministry of Science and Technology, Government of India, New Delhi for the financial assistance.

REFERENCES

Buresh, R.J., M.E Casselman and W.H. Patrick Jr. 1980. Nitrogen fixation in flooded soil systems, a review. *Adv. Agron.*, **23**: 149-192.

De, P.K. 1939. The role of blue-green algae in nitrogen fixation in rice fields. *Proc. Res. Soc. Lond.* **127** B, 121-139.

Gangawane, L.V. and R.S. Saler. 1979. Tolerance of certain fungicides by nitrogen fixing blue-green algae. *Curr. Sci.*, **48(7)**: 306-308.

Goyal, S.K. and G.S. Venkataraman. 1971. Response of high yielding rice varieties to algalization "Interaction of soil types with algal inoculation". *Phykos*, **10:** 32-33.

Gupta, A.B. and A.C. Shukla. 1972. Studies on the nature of algal growth promoting substances and their influence on growth, yield and protein content of rice plants. *Lab dev. J. Sci. Technol.*, 5: 162-163.

Ishizawa, S., and T. Matsuguchi. 1966. Effects of pesticides and herbicides upon microorganisms in soil and water under water logged condition. *Bull. Nat. Inst. Agric. Sci.*, **B:16**: 1-90.

Jagannathan, R. and S. Kannaiyan. 1977. Effects of blue-green algae application on rice yield. *Inte. Rice Res. Newsl.*, **3(4)**: 20.

Kannaiyan, S. 1978. Mass scale multiplication of blue-green algae under field condition. *All India Seminar on blue-green algae and their viruses.* Madurai Kamaraj Univ. (Abst.), p. 32.

Kannaiyan, S. 1979. Blue green algal biofertilizer for rice crop. *Farmers and Parliament*, **14(11)**: 19-25.

Kannaiyan, S. 1980. Effect of certain insecticides on the growth of BGA. p. 492-496. In: *Pesticide residue in the environment in India* (ed.) C.A. Edward, G.K. Veerash and H.r. Krueger , FAO/UNESCO/ICAR/UAS. UAS Tech. Series No 32, p. 523.

Kannaiyan, S. 1981. Nitrogen conservation in rice soil by blue-green algae. *INSFFER Programme Seminar*, Inte. Rice Res. Inst., Manila, Philippines, pp. 1-19.

Kannaiyan, S. 1983. Studies on the effect of algal bio-fertilizer inoculation on rice crop. *National Conf. Society Basic and Appl. Microbiologist,* Dungar College, Bikaner, India.

Kannaiyan, S. 1985. Studies on the algal application for low land rice crop. Tamil Nadu Agric. Univ. Bull. Coimbatore, Tamil Nadu p. 16.

Kannaiyan, S., K. Govindarajan and H.D. Lewin. 1980. Studies on the survival of blue-green algae under dried condition. In: *High lights of Research,* Paddy Experiment Station, Tirur, Tamil Nadu, India. p. 54-59.

Kannaiyan, S., M. Thangaraju and G. Oblisami. 1982. Effect of combined inoculation of blue-green algae with *Azolla* and *Azospirillum* on rice crop. *BNF Newsletter*, **1(2)**: 32.

Koptiyeva, ZN, P. and O.V. Tantsiurenko. 1971. Effect of blue-green algae on growth of rice seedlings. *Microbiol.* Zh. **53(2)**: 215-221.

Kulasooriya, S.A., P.A. Roger, W.L. Barraquio and I. Watanabe. 1980. Biological nitrogen fixation by epiphytic microorganisms in rice fields. *IRRI Res. Pap. Ser.* **47**.

Okuda, A. and M. Yamaguchi. 1955. Nitrogen fixing micro-organisms in paddy soils. I. Characteristics of nitrogen fixation in paddy soils. *Soil and Plant Food.* **1**: 102-104.

Palacpac, A.C. 1982. *World Rice Statistics.* The Inte. Rice Res. Inst., Los Banos, Philippines, 152 p.

Roger, P.A. and S.A. Kulasooriya. 1980. *Blue-green algae and rice.* The Inte. Rice Res. Inst., Los Banos, Philippines, 112 pp.

Roger, P.A. and I. Watanabe. 1985. Technologies for utilizing biological nitrogen fixation in wet land rice, potentialities, current usage and limiting factors. *Fertilizer Review* (in press).

Roger, P.A., S.A. Kulasooriya, W.L. Barraquoo and I. Watanabe. 1979. Epiphytic nitrogen fixation on low land rice. Paper presented at the *workshop on nitrogen cycling in S.E. Asian wet monsoonal ecosystems.* Nov. 5-10, Chiang Mai, Thailand.

Singh, P.K. 1972. Effect of pesticides on blue-green algae. *Proc. 13th Ann. Conf. Assoc. Microbiol., India,* 56.

Subrahmanyan, R. and M.N. Sahay. 1964. Observations on nitrogen fixation by some blue-green algae and remarks on its potentialities in rice culture. *Proc. Indian Acad. Sci. Sect.* B **60(2):** 145-154.

Venkataraman, G.S. 1972. *Algal biofertilizers and rice cultivation.* To-day and Tomarrow's Printers. P. 75.

Venkataraman, G.S. 1979. Algal inoculation in rice fields. In: *Nitrogen and Rice,* Inte. Rice Res. Inst., Philippines.

Watanabe, I. 1978. Biological nitrogen fixation in rice soils. Pages 465-478. In: *Soils and Rice.* The Inte. Rice Res. Inst., Los Banos, Philippines.

Watanabe, I. and S. Brotonegoro. 1981. Paddy fields. Pages 241-263. In: *Nitrogen Fixation,* Vol. I, *Ecology,* ed. W.J. Broughton, Clarendon, Press, Oxford.

Watanabe, I. and T. Kiyohara. 1980. Decomposition of blue-green algae as affected by the action of soil bacteria. *J. Gen. Appl. Microbiol.,* 5(4): 175-179.

Watanabe, I. and P.A. Roger. 1984. Nitrogen fixation in wetland rice field. Pages 237-276. In: *Current Developments in Biological Nitrogen Fixation.* ed. N.S. Subba Rao. Oxford & IBH Publishing Co., New Delhi, Bombay, Calcutta.

Ind. Phyco. Rev. 1993.
2: 83-112.

THE BIOLOGY OF CYANOPHAGES

A.K. KASHYAP, POONAM PRASAD AND P.K. SINGH
Centre of Advanced Study in Botany
Banaras Hindu University, Varanasi - 221 005
and
Central Rice Research Institute
Cuttack, Orissa.

1. INTRODUCTION

Degeneration of algal blooms in nature without any apparent reasons had created considerable controversies among scientists in order to explain this phenomenon. It was Safferman and Morris (1963) who succeeded in isolation of virus infective on blue-green algae (cyanobacteria) by employing the techniques of microbiology. This pioneering investigation partially explained the problem of algal degeneration in natural water bodies. Subsequent investigations by group of workers from Israel (Padan *et al.* 1967), India (Singh and Singh 1967a), Japan (Ueda 1965), Sweden (Granhall and Hofsten 1969), Russia (Goryushin and Chaplinskaya 1966), Australia (Franche 1987)and America (Safferman *et al.* 1969a, b; Adolph and Haselkorn 1973a, b; Hu *et al.* 1981) made it amply clear that viruses of cyanobacteria have widespread geographical distribution.

The isolation of cyanophages attracted attention of physiologists, biochemists, molecular biologists and virologists because studies on virus infection made on higher plants were severely hampered due to complexity of host-tissue-systems. In contrast cyanobacteria provided eukaryotic type of photosynthesis combined with prokaryotic cellular organization and they could be handled like any other microorganism. Thus cyanobacterial-virus system was adopted as a model for studying host-virus interactions during subsequent periods to come. The discovery of blue-green algal viruses was of considerable interest to those who were working on cyanobacterial metabolism, photosynthesis and genetics. Especially those interested in genetics thought to utilize the viruses, much the same way as bacteriophages were used, for genetic analysis of cyanobacteria through transduction experiments. On the other hand, most of the ecologists viewed cyanobacterial viruses as a potential tool to control undesirable growth of cyanobacterial blooms in freshwater bodies. The subject has been reviewed by various workers in the field (Brown 1972; Padan and Shilo 1973; Safferman 1973; Sherman and Brown 1978; Singh and Padhy 1982; Kaushik and Venkataraman 1982; Gromov 1983).

2. CYANOPHAGE CHARACTERISTICS

(a) *Typology*

Safferman and Morris (1963) first isolated a virus which destroyed blue-green algae. Since then quite a number of viruses were discovered attacking blue-green algae and have been designated as phycovirus (Safferman *et al.* 1969b), algophages (Goryushin and Chaplinskaya 1968), blue-green algal viruses (Safferman and Morris 1963) and cyanophages (Luftig and Haselkorn 1967). The term 'cyanophages' has been preferred over other terms since there has been a general concensus to designate blue-green algae as cyanobacteria (Stanier and Cohen-Bazire 1977) and will be used hereafter.

The nomenclature of cyanophages has been much more systematic than of bacteriophages which have been based on choice of an investigator(s). Although the nomenclature of bacteriophages appears to be haphazard, it has served very well for day to day practice. In contrast cyanophages have been named by the initials of generic names of hosts. They were further grouped on the basis of serological specificities and morphological properties (Padan and Shilo 1973). Serologically different subgroups have been designated by numerical system (Safferman *et al.* 1969a) and are arranged in the chronological order of their isolation (Ginzburg *et al.* 1968; Safferman *et al.* 1969a; Daft *et al.* 1970). Viruses infecting identical strains of *Lyngbya, Plectonema* and *Phormidium* were placed in group LPP. Out of eleven strains (Safferman and Morris 1967) five possessed single viral form LPP-1 (Safferman and Morris 1963), whereas, the other six were serologically different and were designated as LPP-2 group (Safferman *et al.* 1969a). Cyanophages SM, N and AS groups has only one virus form, therefore, designated as SM-1, N-1 and AS-1, respectively. The viruses A-1, A-1(L), A-2 and A-4(L) are confusing as all of them infect *Anabaena.* Moreover, taxonomic classification of blue-green algae proved misleading for *Anacystis, Synechococcus* and *Microcystis*, therefore, stress was given on the serological properties of virus and their grouping.

(b) *Source and host range*

Cyanophages have been isolated from waste stabilization ponds, drains, sewage oxidation ponds and rice fields. Among these highest titre of viruses were recorded in waste stabilization ponds (Safferman and Morris 1964b, 1967; Padan and Shilo 1969; Shilo 1971,1972). Viruses, their source of isolation and host range has been summarized in Table 1. LPP-viruses, in addition to pond of Indiana, were subsequently isolated from fish ponds and lakes (Padan *et al.* 1967; Singh and Singh 1967a; Padan and Shilo 1969), Delaware rivers of USA (Shane 1971) and even from soil (Jackson and Sladecek 1970). Existence of cyanophages has also been reported in fresh water habitats (Goryushin and Chaplinskaya 1966, 1968).

LPP-1 virus attacks morphologically similar members of the non-heterocystous filamentous genera *Plectonema, Phormidium* and *Lyngbya* (Safferman and Morris

Table 1. Blue-green algal viruses, their source and host range.

Virus	Source	Host range	References
AS-1	Waste stabiliz-ation pond, Florida	*Anacystis nidulans, Synechococcus cedrorum*	Safferman *et al.* (1972)
AS-1M	---"---	*A. nidulans, S. cedrorum*	Sherman & Connelly (1976)
AS-2	---"---	*A. nidulans, S. cedrorum, Synechococcus* sp. (NRC-1)	Stewart & Daft (1976)
AM-1	---"---	*A. nidulans*	Karbysheva *et al.* (1974) Vorontsova *et al.* (1974)
S-1	---"---	*Synechococcus* strain NRC-1	Adolf & Haselkorn (1973b)
SM-1	Waste stabilization pond, Indiana	*Synechococcus elongatus, Microcystis aeruginosa*	---"---
SM-2	---"---	---"---	Safferman & Morris (1967)
LPP-1	---"---	*Lyngbya, Plectonema, Phormedium*	Safferman & Morris (1963)
LPP-1D	---"---	---"---	Cannon *et al.* (1971)
LPP-1 (D1)	---"---	---"---	Daft *et. al.* (1970)
LPP-1G (III)	Fish Pond, Israel Polluted Water, India	---"---	Padan *et al.* (1967) Singh & Singh (1967a)
LPP-1M	---"---	---"---	Adolph & Haselkorn (1972b)
LPP-2	Waste stabilization pond Indiana Florida, Missouri and Arkansa	---"---	Safferman *et al.* (1969a)
LPP-2SPI	---"---	*Lyngbya, Plectonema, Phormidium*	Padan *et al.* (1972)
LPP-2N	---"---	---"---	Booth (1975)

Virus	Source	Host range	References
LPP-3	---"---	---"---	Mendzhul *et al.* (1975)
P-2	Polluted water, India	*Plectonema boryanum*	Singh and Singh (1967a)
P-3	---"---	---"---	Singh & Singh (1967a)
P-4	---"---	---"---	Singh and Singh (1967a)
P-5	---"---	---"---	Singh *et al.* (1972)
Long tailed virus	---"---	---"---	Singh (1973)
P-6	---"---	*P. boryanum, Anabaenopsis reciborskii, R. indica*	Singh *et al.* (1972)
N-1	Fresh water lake, Wisconsin	*Nostoc muscorum, Anabaena variabilis*	Adolph and Haselkorn (1971) Hu et al. (1981)
N(S)1	Rice field, Senegal	*Anabaena strain* 77S15 *Nostoc muscorum*	Franche (1987)
A-1	Polluted waters, India	*Anabaena cylindrica*	Singh *et al.* (1972)
A-2	Soil, Sweden	*Anabaena variabilis*	Granhall and Hofsten (1969)
A-1(L)		---"---	Kozyakov *et al.* (1972)
A-4(L)		---"---	Kozyakov and Gromov (1973)
AR-1	Polluted water, India	*Anabaenopsis raciborskii, A. circularis, Raphidiopsis indica*	Singh and Singh (1967a)
AR-2	Polluted water, India	*A. raciborskii*	Singh *et al.* (1972)
An11-AN26	Sewage settling ponds	*Anabaena, Nostoc*	Hu *et al.* (1981)
C-1	Polluted water, India	*Cylindrospermum* sp.	Singh and Singh (1967a)

1964a). Base ratio (46% G+C) and density (1.705 g/cm^3) of these genera are very similar (Edelman *et al.* 1967). Cyanophage SM-1 attacks unicellular *Synechococcus elongatus* (Safferman *et al.* 1969b) whose G+C% (66-67%) is higher than filamentous organisms (Mackenzie and Haselkorn 1972a) while density is all the more same (1.73% g/cm^3). AR-1 and AR-2 attacks *Anabaenopsis raciborskii, Anabaenopsis circularis* and *Raphidiopsis indica* (Singh and Singh 1967a). SM-2 infects *Synechococcus elongatus* and *Microcystis aeruginosa* (Safferman and Morris 1967). Series of viruses P-2 to P-6 were reported attacking *P. boryanum* (Singh and Singh 1967a; Singh *et al.* 1972; Singh 1973). Recently some new *Anabaena* and *Nostoc* cyanophages (AN-10, 15, 20, 22, 24) from sewage settling ponds have been reported from Senegal (Franche 1987) and rice field (Hu *et al.* 1981).

(c) *Physico-chemical properties*

(i) *Morphology*

Cyanophages resemble T-even or T-odd coliphages in their general morphology. Head of cyanophages is mostly hexagonal, icosahedral or polygonal in shape with a short or long tail (Table 2). Cyanophages with noncontractile tails (LPP type) resembled T-7 coliphages (Adolph and Haselkorn 1972b; Safferman *et al.* 1972; Luftig and Haselkorn 1967), while those, with contractile tails (N-1 and AS-1) resembled T-even coliphages (Adolph and Haselkorn 1971; Safferman *et al.* 1972). Besides, cyanophage belonging to LPP group exhibiting long contractile tail has also been reported (Singh 1974). Tail is, mostly, a hollow cylinder of usually 2-4 rings of six subunits in 6 fold radial symmetry, joined to head through tail capital. Spikes and fibres are also reported to be present at the distal end of tail. In SM-1 short collar with thin appendages is present (Mackenzie and Haselkorn 1972a, b) and a tail with base plate and short pins has been reported for AS-1 (Kirilova and Chaplinskaya 1973). Head is formed of capsomers and model for LPP-1 has been presented (Padan and Shilo 1973).

(ii) *Composition*

All the cyanophages (LPP-1, AS-1, S-1 and SM-1) possess linear double stranded DNA as genetic material, which is in confirmation with Bradley's (1967) prediction that head tail bacteriophages contains only double stranded DNA. Schneider *et al.* (1964) first showed that LPP-1 contains DNA, later its physical and chemical properties were investigated in detail (Goldstein *et al.* 1967; Luftig and Haselkorn 1967; Sherman and Haselkorn 1970a, b). Subsequently DNA of few other viruses were analysed and their properties have been summarized in Table 3.

Structural proteins were characterized by sodium dodecyl sulfate gel electrophoresis of bulk viral proteins and separated head tail proteins (Luftig and Haselkorn 1968a; Sherman and Haselkorn 1970c). LPP-1 consists of 10 principal structural proteins of molecular weight 900,000 daltons. Out of this the major head

Table 2. Morphology of blue-green algal viruses.

Virus	Shape of Head capsid	Edge to edge diam. of capsid (nm)	Tail size lxb (nm)	Tail properties	Reference
A-1(L)	Icosahedral	60	82.5	Long	Kozyakov (1974)
AN-10	Polyhedral	60±4		Long contr-actile with base plate and pins	Hu *et al.* (1981)
AN-15		63±4		"	Hu *et al.* (1981)
AN-20		51±2		"	Hu *et al.* (1981)
AN-22		54±3		Short tail	Hu *et al.* (1981)
AN-24		50±1		"	Hu *et al.* (1981)
A-2	Octahedral	63	100 (l)	Long, contractile	Mendzhul *et al.* (1975)
A-4(L)	Polyhedral	56-62	--"--	Short	Khudiakov & Gromov (1973)
AS-1	Hexagonal	90	243.5x22.5	Long, rigid, contractile base	Safferman *et al.* (1972)
			88-120(l)	Plate with short pins	Kirillova & Chaplin-skaya (1973).
AS-1M	Hexagonal	90	240x22	Rigid contractile	Sherman & Connelly (1976)
S-1	Hexagonal	50	140 (l)	Long, narrow, rigid, noncont-ractile	Safferman et al. (1969b)
S-2L	Icosahedral	56	120 (l)	Flexible, non-contractile	Khudyakov (1977)
SM-1	Icosahedral	67-88	Ab	Short collar with thin appendage	Mackenzie & Hasel-korn (1972a)
SM-2	Polyhedral	50-55	130-140(l)	Long, narrow, felxible noncont-ractile	Fox *et al.* (1976)

Virus	Shape of Head capsid	Edge to edge diam. of capsid (nm)	Tail size lxb (nm)	Tail properties	Reference
LPP-1	Icosahedral	58.6$\pm$2.1	20x15	Short, stubby, noncontractile	Luftig & Haselkorn (1967) Mendzhul & Zhigir (1971)
Long tailed virus	Polyhedral	54.5	116x16	Long	Singh (1974)
N-1	Hexagonal	55 to 61.4 $\pm$2.8	110 x 16	Rigid, contractile striated flexible fibre at neck	Adolph & Haselkorn (1971, 1973a)
N(S)1	Hexagonal	50	10	Short, rarely visible	Franche (1987)

Table 3. Physicochemical properties of cyanophage DNA.

Virus	Mol wt $x10^6$	Contour length (nm)	Buoyant density CsCl (g/ml)	Sedimentation coefficient ($S_{20}W$)	Melting temp.	G + C%
AS-1M[1]	57$\pm$3	29.5	1.71 at 20 ^{0}C	45.6$\pm$0	74	52-55
S-1[2]	23-26	13.3$\pm$0.3	1.729	30.8$\pm$1.5	82	70-74
SM-1[3]	56-62	24.3	1.725	48.0	88-92	66-67
LPP-1[4]	25.5-27	13.2 u (11-15.5 u)	1.714 at 25 ^{0}C	33.4 35.12b	--	53-55
N-1[5]	38-45	17.2 u	1.696	40.9$\pm$1.4	--	37-41
N(S)1[6]	28 (Kb)	--	--	--	--	--

References : 1. Sherman and Connelly (1976)
2. Adolph and Haselkorn (1973b)
3. Safferman *et al.* (1969a), Mackenzie and Haselkorn (1972b)
4. Goldstein and Bendet (1967)
5. Luftig and Haselkorn (1967)
6. Franche (1987)
b. Moskovets *et al.* (1971)

proteins weighed 39,000 and 13,000 daltons while the main tail component was constituted by 80,000 daltons molecular weight species (Sherman and Haselkorn 1970c). LPP-2 differed from LPP-1 in having 14 structural proteins of 749,000 daltons mol wt species (Sherman 1970).

Mature SM-1 contains only a dozen structural proteins of which two (40,000 and 25,000 mol wt) are the major components of capsid shell, containing 80% sulfur atoms (Mackenzie and Haselkorn 1972a). Comparatively N-1 contained maximum structural proteins (19) of 59,000 mol wt, out of this, major structural protein weighs 37,000 and 14,000 daltons. Structural proteins accounted for 16%, 34-35% and 50% of DNA of SM-1, N-1, LPP-1 and LPP-2 respectively (Sherman 1970; Sherman and Haselkorn 1970c; Adolph and Haselkorn 1971; Mackenzie and Haselkorn 1972a).

(iii) *Temperature and pH sensitivity*

The most important environmental factors affecting distribution of cyanobacteria and their viruses are temperature and pH. Effect of temperature on phage stability is extensively studied and their results are summarized in Table 4. Although the greatest stability of phages were observed at 40-45^{0}C, they could tolerate even 55-60^{0}C, with the exception of A-1 which tolerated only 28-30^{0}C (Mendzhul *et al.* 1974). Survival of phages at particular temperature is directly correlated with the incubation period. Virus particles of P-5 and P-6 requires only 19 and 2.5 min preincubation (Singh *et al.* 1972).

Table 4.Effect of temperature on cyanophage survival.

Virus	Temperature (^{0}C)	Survival (%)	Reference
AS-1	45	100	Safferman *et al.* (1972)
	50	77	
	55	12	
	60	0.001	
AS-1M	45	100	Sherman & Connelly (1976)
	50	80	
	55	10	
SM-1	40	98	Safferman *et al.* (1969b)
	45	35	
	55	0.001	
SM-2	40	Stable	Fox *et al.* (1976)
	60	lost	
LPP-1	55	0.001	Safferman & Morris(1964a)
LPP-(D1)	60[a]	Inactivated	Daft *et al.* (1970)

Virus	Temperature (^0C)	Survival (%)	Reference
LPP-2N	45	Stable	Booth (1975)
	55	Inactivated	
N-1	45	Stable	Padhy & Singh (1977a)
	50	62	
	55b	10	
	60	Inactivated	
N(S)1	65	Stable	Franche (1987)
	70^c	40	
	75^b	Complete inactivation	

% survival was measured after 1 hr preincubation

a = 6 min incubation
b = 20 min incubation
c = 60 min incubation

All the phages are stable in the pH range 5-11, with the exception of N-1, A-1 and A-1L which are sensitive to acidic pH 7.0 (Kozyakov *et al.* 1972; Mendzhul *et al.* 1974; Singh and Padhy 1982). Both AS-1 and AS-1M lost 75 and 50% activity at pH 11.0 and 5 log units at pH 3.0 (Safferman *et al.* 1972; Sherman and Connelly 1976). The ranges of pH (5-11) for cyanophages survival accords with the pH range of host suggesting that phages can significantly control algal blooms.

(iv) *Effect of metallic ions*

Requirement of Mg^{2+} for phage stability and high multiplicity is reported in LPP-1 and N-1 (Table 5; Adolph and Haselkorn 1972a, b; Padan and Shilo 1973; Padhy and Singh 1978a) whereas SM-1, AS-1, AS-1M and A-1L, N(S)1 are stable in distilled water (Safferman *et al.* 1969a; Kozyakov *et al.* 1972; Safferman *et al.* 1972; Sherman and Connelly 1976; Franche 1987). Further it was concluded that requirement of divalent cation (Mg^{2+}) is possibly properties of viruses infecting filamentous blue-green algae. Mg^{2+}, Ca^{2+} (alkaline earth metals) and Zn^{2+} (transition metal) are reported to be extensively involved in protein structure (Vallee and Walker 1970; Von Hippel and McGhee 1972; Lijas and Rossman 1974). Possibly Mg^{2+} is involved in secondary and tertiary coiling of peptide chain of viral coat protein as in its absence viral capsid is disintegrated followed by liberation of nucleic acid (Luftig and Haselkorn 1967;

Table 5. Physicochemical properties of cyanophage particles.

Virus	Sedimentation coefficient	Buoyant density (g/cm^3)	pH range of stability	Mg^{2+} ment (mm)	References
A-1	-	-	7-10.5	-	Mendzhul *et al.* (1974)
A-1(L)	-	-	7-10	NR	Kozyakov *et al.* (1972)
AS-1	-	-	4-10	NR	Safferman *et al.* (1972)
AS-1M	754	1.49	4-10	NR	Sherman and Connelly (1976)
SM-1	1.021	1.48	5-11	NR	Safferman *et al.* (1969a)
SM-2	-	-	5-11	-	Fox *et al.* (1976)
LPP-1	555-548	1.48	5-11	1.0	Padan and Shilo (1973)
N-1	539	1.498	6.7-10.5	2.0	Adolph and Haselkorn (1971, 1973a) Padhy and Singh (1977a, 1978a)
Long tailed virus	-	-	-	2.0	Singh (1974)
N(S)-1	-	-	5.0-11.0	NR	Franche (1987)

Goldstein *et al.* 1967). Role of ions can further be ascertained by employment of chelating agent, EDTA. It binds with cations bound to virus particles (Lark and Adams 1953) and inactivates cyanophages (Luftig and Haselkorn 1968a; Padan *et al.* 1972; Singh 1974, 1975; Padhy and Singh 1977a, c) resulting in separation of tails and bursting of head releasing DNA (Luftig and Haselkorn 1968a). Complete inactivation of N-1 occurred at 5-100 mM of saline solution (Padhy and Singh 1978a) however, its requirement is reported in AS-1 (Safferman et al. 1972). If Na^+ of saline $MgCl_2$ is replaced by K^+ (100 mM), the stabilizing efficiency decreases (Padhy and Singh 1978a).

3. GROWTH CYCLE OF CYANOPHAGES

(a) *Adsorption*

Cyanobacteria is organized as a typical prokaryote. Following cell wall is a cell membrane which is typical of gram negative bacteria. Nucleoplasm having nucleic acid is separated from cell membrane by peripheral photosynthetic lamellae (thylakoid) as in higher plants. In addition it contains polyphosphate bodies, glycogen, polyhedral bodies, α - granules, β - granules and cyanophycin granules (Smith *et al.* 1966a, b; Lang 1968; Chao and Bowen 1971; Simon 1971; Whitton *et al.* 1971).

Cyanophage adsorbs to the host cell at specific receptor sites (Padan *et al.* 1967) by the distal end of its contractile tail (Smith *et al.* 1967). It penetrates the wall by forming ionic bonds (Smith *et al.* 1966a,b, 1967) and through the contraction of tail sheath and penetration of core injects its DNA into the host cell (Gromov *et al.* 1974). Details of infection is not known, however, for AS-1M it is reported that phage is attached to the host cell by tail pins (Sherman *et al.* 1976).

The adsorption rate constants of cyanophages is 3-5 x 10^{-11} cm^3/sec (Adolph and Haselkorn 1972a; Padan and Shilo 1973) and most of them follow first order kinetics, however, phage and host cell ratio affects adsorption rate in LPP-1 (Goldstein *et al.* 1967; Safferman *et al.* 1972). Adsorption occurs in two steps (1) A physical and reversible process, and (2) Physiological and irreversible process, wherein chemical bonding results in stable attachment. The process is significantly affected by the factors such as host ageing, temperature, pH and ions (Padhy and Singh 1977b, 1978b). Cultures of stationary phase shows altered metabolic status which eventually affects the adsorption rate of N-1 (Padhy and Singh 1977b). The energy of activation for the process is found to be 16.3 Kcal/mole. Exposure of cells above critical temperature is known to denature membrane protein irreversibly and consequently the adsorption rate. Unlike bacteriophages the N-1 adsorption was sharply affected by pH below 7.6 and above 8.1 (Padhy and Singh 1978b) suggesting involvement of amino groups. With the addition of NaCl (100 mM) the adsorption rate of AS-1 to *Synechococcus cedrorum* and AS-1M is increased by 25 and 5%, respectively (Safferman *et al.* 1972; Sherman and Conelly 1976). Sodium citrate (10mM) inhibited N(S)1 attachment to host (Franche 1987).

In comparison to bacteriophages, relatively little information is available about the cyanophage receptor complexes present on the cell surface. In this regard studies conducted by Samimi and Drews (1978) is significant as they isolated the receptor complex of cyanophage AS-1 from unicellular cyanobacterium *Anacystis nidulans* KM. Chemically receptor is made up of proteins ranging from 54,000 to 64,000, whereas lipopolysaccharide(s) of cell wall are responsible for first steps in recognition.

(b) *One step growth*

Following adsorption of the cyanophage on the host cell, growth of virus ensures showing sequentially latent period, eclipse period, rise period and plateux. Though this pattern is remarkably similar to those exhibited by bacteriophages, it proceeds at comparatively slower rate, possibly because their photoautotrophic nature renders them as slow grower. The characteristic features of growth cycle of cyanophages has been summarized in Table 6. Cyanophages show great variation in the extent of latent period which last just for about 2-30 h, averaging from 5-7 h. Likewise the rise period lasted from 3-20 h.

Great variation in number of plaque forming units (PFU) liberated following infection of a cell (burst size) is much dependent upon cultivation conditions of the host and incubation during growth cycle of the virus.

The burst size on the basis of the one step growth experiments ranges from 40-350 PFU/infected cell in various cyanophages. The situation is more uncertain in the case of cyanophages which infect filamentous cyanobacteria.

The elipse period for each of the viruses has not been determined critically but where known with certainty has been summarized in Table 6.

(c) *Ultrastructural changes after viral infection*

Cyanophage infection is characterized by invagination of photosynthetic lamellae (Brown *et al.* 1966; Smith *et al.* 1966b) followed by formation of a space, virogenic stroma, between thylakoid and cell membrane (Smith *et al.* 1966b). As soon as phage DNA reaches nucleoplasm protein synthesis of host cell ceases with subsequent replication of phage DNA. Replicated DNA forms compound helices of 180^{0}A diametre and finally reaches to the periphery forming virogenic stroma. Presence of phage antigens in virogenic stroma suggests it to be as the centre of phage assembly. In LPP-1 possible damage to thylakoid is reflected as presence of ribosomes and glycogen in virogenic stroma. Moreover requirement of host DNA and viral proteins for thylakoid invagination is reported (Sherman and Haselkorn 1970a, b). By the end of latent period (LPP-1, 7 h) most of the host cells shows presence of viral heads and by 10th h phage assembly starts (Sherman and Haselkorn 1970a). Heterocysts and spores are rather phage resistant (Singh and Singh 1967a; Granhall and Hofsten 1969).

Table 6. One step growth of cyanophages.

Virus	Latent period (hr)	Rise period (hr)	Burst size PFU/infected cells	References
A-1(L)	5	-	75-75	Kozyakov (1974)
A-4(L)	2.5	-	100-120	Khudiakova and Gromov (1973)
A-1(L)	5	8	110	Hu *et al.* (1981)
AN-10	6	8	150	Hu *et al.* (1981)
AN-15	5	7	175	Hu *et al.* (1981)
AN-20	1	3	75	Hu *et al.* (1981)
AN-22	2	2.5	30	Hu *et al.* (1981)
AN-24	2	4	30	Hu *et al.* (1981)
AS-1	8.5	7.5	50	Safferman *et al.* (1972)
AS-1M	8	4	40	Sherman *et al.* (1976)
S-2L	5	-	100	Kozyakov *et al.* (1972)
SM-1	30	20	100	Mackenzie and Haselkorn (1972b)
LPP-1	7	3	100	Goldstein *et al.* (1967) Sherman and Haselkorn (1970)
LPP-1G	C-6.5	6	350	Padan *et al.* (1970)
Long tailed	2	-	100	Singh (1974)
N-1	7	6	100-300	Adolph and Haselkorn (1972a)
N(S)1	20	28	70	Franche (1987)

In contrast to LPP-1, phages (SM-1, A-1L, S-2L) infecting unicellular organisms do not show change in photosynthetic lamellae prior to lysis (Gromov *et al.* 1974; Khudiakov 1977) and nucleoplasm is the only centre of virion replication and accumulation. Similarly, AS-1 infection does not show any gross morphological change in host cells prior to lysis (Pearson *et al.* 1975) except accumulation of electron dense material in nucleoplasm which is possibly precursor of virus particles. Just prior to lysis virus particles gets condensed around thylakoids followed by its disruption. Comparatively AS-1M infection was characterized by (1) presence of electron transparent nucleoplasm due to degradation of host DNA, (2) increase in fibrillar material due to phage DNA synthesis, (3) breaks in thylakoid instead of invagination, and (4) formation of virogenic stroma together by centroplasm and chromatoplasm (Sherman *et al.* 1976). After lysis cell wall and cell membrane forms vesicles as a result of action of phage lysozyme (Pearson *et al.* 1975).

(d) *Synthesis of viral components and assembly*

Biosynthesis of DNA, funotional and structural proteins and production of virions have been extensively studied in LPP-1 and reviewed by Padan and Shilo (1973). Virus directed DNAse degrades 50% of the host DNA to acid soluble product within 7 h of infection, which is reincorporated into viral DNA (Sherman and Haselkorn 1970c). The pattern of protein synthesis of bacteriophage T7 and LPP group has much similarity (Sherman and Haselkorn 1970c), however, the latter differs in utilizing additional second major structural protein in building the viral shell (Adolph and Haselkorn 1972b).

Soon after cyanophage infection host protein synthesis is depressed and virus induced fresh protein synthesis starts. Three classes of major proteins are known (1) earliest class, (2) early class and (3) late class. The first two class consists of three proteins each whose synthesis starts from 0-4 h and 2 h of lysis, respectively. However, late class proteins (structural proteins) are synthesized from 4 h until lysis. In all 17 such proteins are synthesized accounting for 65% coding capacity of LPP genome.

Likewise, AS-1M infection to *Synechococcus cedrorum* stopped hosts protein synthesis after 1-2 h and by 4 h of infection host DNA was completely degraded into acid soluble forms which was reincorporated into phage DNA by the end of latent period (3 h). Early and middle proteins were synthesized at 2-4 h and 3 h lysis, respectively. Late proteins were synthesized simultaneously with that of middle proteins (Sherman and Pauw 1976).

(e) *Cyanophage infection and host-metabolism*

Electron microscopic observations on LPP-1 infected cells of *P. boryanum* undoubtedly showed that thylakoid membranes are altered and therefore it was anticipated that it might affect photosynthesis. Considerable amount of works were conducted in this direction. It was demonstrated that LPP-1 infection had no influence on O_2 evolution (Wu and Shugarman 1967), but completely arrested CO_2 photoassimilation (Ginzburg *et al.* 1968). This implied that LPP-1 replication was partially dependent upon photophosphorylation since no change in photosynthetic machinery was observed (Ginzburg *et al.* 1976) and that CO_2 fixation was not required for virus replication. Thus virus synthesis (10%) was observed even in complete darkness while CCCP and anaerobiosis in dark completely arrested the productive cycle (Padan et al. 1971; Padan and Shilo 1973). These observations suggested that energy for viral replication was derived from cyclic phosphorylation and that complete functioning of photosystems was necessary. This was further supported by the fact that withdrawal of light affected the latent period as well as burst size (Padan and Shilo 1973).

In contrast to LPP-1, development of SM-1 and AS-1 was more dependent upon host metabolism as light reactions and CO_2 fixation remained unaltered until lysis (Mackenzie and Haselkorn 1972b; Allen and Hutchinson 1976). In case of AS-1M, O_2 evolution as well as PS II activity was insignificantly affected (Sherman 1976). Studies on N-1 infected *N. muscorum* cells showed that there was a change in photosynthetic pigments which was more pronounced at 4 h of infection. There was a significant loss in phycobilin pigments (Kashyap and Singh 1989) which was associated with a concomitant decrease in O_2-evolution (84%) but not observed by Amla *et al.* (1987). Measurement of PS II dependent electron transport activity also showed a decrease (75-80%) in contrast to uninfected cells (Kashyap and Singh 1989), loss of phycobilins has been considered as an indicative of nitrogen stress in the cells (Allen and Smith 1969). Although not verified experimentally, it seems that virus infection induces phycobilin degradating enzymes. In spite of the fact that PS II activity decreased during the latent period, the electron transport activity of PS.I was retained (Kashyap and Singh 1989). Declined PS II mediated electron transport activity following infection may explain the apparent failure of DCMU to inhibit viral replication described by Adolph and Haselkorn (1972a). Since cellular ATP pool declined to a level of 30% (Kashyap and Singh 1989) and $^{14}CO_2$ incorporation decreased to 60% (Adolph and Haselkorn 1972a), question arises regarding the source of energy. This could be the respiratory pathway which was confirmed by demonstration of increased glucose-6-phosphate dehydrogenase activity, decrease in carbohydrate reserve and increased Mg^{2+} dependent ATPase activity (80%) than the Ca^{2+}-ATPase activity (Kashyap and Singh 1989). Activation of oxidative pentose phosphate pathway has also been documented for AS-1 infected cells of *Anacystis nidulans* (Balogh *et al.* 1979) and N-1 infected cells of *Nostoc muscorum* (Amla *et al.* 1987).

Little is known about the effect of virus multiplication on the nitrogen metabolism of a cyanobacterium. The only system which has been tested in this regard pertains to *Nostoc muscorum* and cyanophage N-1. An elaborate relationship between heterocysts and vegetative cells has been reviewed (Haselkorn 1978; Stewart 1980) and no one doubts nowadays that heterocysts are the sites of nitrogen fixation. The impact of cyanophage N-1 development on nitrogenase, glutamine synthetase (GS) and aminotransferase [glutamate oxaloacetate transaminase (GOT); glutamate pyruvate transaminase (GPT)] activities was followed during its latent period (Kashyap *et al.* 1988). The nitrogenase activity was inhibited after 2 h of infection suggesting that phage development does not require the product of nitrogenase activity (Kashyap *et al.* 1988). GS activity was not inhibited until 4 h of infection, however, a decline in activity was subsequently observed. GOT was inhibited after 1 h of infection and no activity was detectable during the entire latent period. In contrast GPT activity increased 2-fold by 4 h of infection and remained higher than the background level until the end of latent period. It was concluded that under N_2-fixing conditions, N-1 multiplication proceeds in absence of N_2-fixation and the amino acid metabolism was altered, probably to meet the requirement of specific amino acids involved in phage directed protein synthesis (Kashyap *et al.* 1988). The decline in nitrogenase activity could be due to inhibition of uptake hydrogenase activity and led to an unmasking of H_2 evolution. The increased H_2 evolution could be due to activity of reversible hydrogenase or ATP dependent hydrogenase activity (nitrogenase) (Kashyap *et al.* 1988).

Nitrate reductase activity was adversely affected following N-1 infection, possibly due to decrease in nitrate reductase apoprotein by viral induced specific proteases (Bagchi *et al.* 1987).

4. LYSOGENY

Although existence of lysogeny was suspected in cyanobacteria and claims to isolate such strains were made (Cannon *et al.* 1971; Cannon and Shane, 1972), the data supporting these assertions were far less satisfactory and were not subsequently followed. More rigorous proofs of lysogeny were forwarded for the cyanophage LPP-2 SPI which was initially isolated by Safferman *et al.* (1969b) and belongs to LPP-2 class. After a careful study, Padan *et al.* (1972) isolated a clone from the centre of a turbid plaque which they referred to as 'SPI'. This strain was similar to *Plectonema* (wild type) but differed in that (a) over 90% of the population was capable of producing phage, and (b) the strain was immune to superinfection by LPP-2 but not to cyanophage LPP-1 group. However, the authors were unable to demonstrate induction with UV light, X-rays or mitomycin-C.

The problem of induction was overcome by isolation of temperature sensitive mutant of LPP-2 SPI forming clear plaque (SPI *cst* 1) which lysogenized *Plectonema* in a stable fashion at 26^{0}C. The induction took place with the formation of progeny phage if the lysogenized cells were exposed to 40^{0}C. The kinetics of induction showed that intracellular phage was released after one hour of incubation at 40^{0}C and lysis was complete within three hours after induction. Silverberg *et al.* (1977) followed the development of cyanophage after induction through electron microscopy. They showed that upon induction viral replication is always found in the nucleoplasm and invagination of the photosynthetic lamellae does not take place. This is in contrast to the observations made for LPP-1.

Rimon and Oppenheim (1975) investigated the physiological requirements for induction. They showed that the optimal temperature for induction was 40^0C which requires at least one hour of simultaneous exposure to light. Such a combination of treatment is necessary for an early step in the induction process. Although CO_2 photoassimilation continues unaffected after induction until the onset of lysis, induction may proceed in absence of CO_2 . DCMU treatment at or before the heat pulse caused inhibition of induction (less than 1% of the control level) while CCCP completely prevented induction.

The overall pattern of induction appears to be similar to those of bacteriophage. The experiments with heat induction have suggested that mutated genes possibly code for temperature sensitive protein whose function is to repress prophage development and in this respect it is likely to resemble the CI gene product of bacteriophage pronounced a 'lamda'. However, light seems to play a crucial role in that it may be involved in the repressor inactivation, or early gene expression or excision of prophage.

Franche (1987) has described occurrence of lysogeny in *Anabaena* 77S15 following infection with cyanophage N(S)1 and two stable lysogenic clones 84SL1 and 84SL2 were isolated. The development of prophage could be induced with mitomycin-C as in the case of N-1 in a lysogen of *Nostoc muscorum* ISU (*Anabaena* PCC 7120) (Padhy and Singh 1978d). Failure to detect extrachromosomal DNA of a molecular weight equivalent to that of cyanophage N(S)1 (i.e. 28 kb) in the lysogenic strains of *Anabaena* 77S15, it was concluded that the prophage was integrated into the host chromosome.

5. GENETICS

(a) *Mutation in the hosts*

Various types of mutants of *P. boryanum* has been described which include resistant mutants (cyanophage LPP-1 and streptomycin) and short trichome mutants. The spontaneous mutation to virus resistance occurred with a frequency of 1.0 - 2.5 x 10^{-7} (Padan *et al.* 1967; Gromov and Kozyakov 1970; Singh *et al.* 1972; Singh and Kashyap 1977). Induction of mutations for LPP-1 and streptomycin resistance by various mutagenic agents have been investigated and their efficiency in inducing virus resistance were in the order: MNNG >UV >acriflavine> 2AP >caffiene (Singh and Kashyap 1977). The spontaneous and MNNG induced N-1 resistant mutants of *N. muscorum* has been also documented with average mutation frequenccy of 3.55 x 10^{-5} and 6.08 x 10^{-4}, respectively (Padhy and Singh 1976). The reversion frequency of N-1 resistant population was reported to be 3.99 x 10^{-6}/cell/generation (Padhy and Singh 1978c).

(b) *Mutations in cyanophages*

Safferman and Morris (1964a) were the first to describe two types of variant plaque morphologies for LPP-1 and each of them reverted to a mixture of both. Subsequently mutants of LPP-1, LPP-2 SPI and AS-1 were described by various workers and has been summarized in Table 7. Most of the mutants were found to be stable and each of the genetic markers were induced by a range of mutagenic agents. The rapid lysis (*r*) mutants of LPP-1 were found to have a shorter latent period than the wild type (r^+) (Singh *et al.* 1972). On the other hand host-range (*h*) mutants had latent period similar to wild type (h^+). Various *h* mutants could be further distinguished on the basis of plaque morphology, efficiency of plating on LPP-1 resistant mutant and spontaneous and induced reversion frequencies (Kashyap 1980). The temperature-sensitive (*ts*) mutants of LPP-1 were unable to grow at non-permissive temperature (40^{0}C) and the temperature sensitive phase lasted for 2-3 h during intracellular growth as judged by shift-up and shift-down experiments. The mutants differed from the wild-type (ts^+) in being more sensitive to photodynamic inactivation and chelating agent (EDTA) shock. They exhibited high frequency towards *r* mutation following exposure of free phage particles to high temperature (Singh and Kashyap 1976). Based on mutagenic specificity it has been suggested that *r* mutants originated as a result of frame shift mutantions whereas *h* and *ts* mutant arose due to point mutations (transitions transversions) (Singh and Kashyap 1976).

Rimon and Oppenheim (1974) isolated 23 *ts* mutants of the temperature cyanophage LPP-2 SPI following treatment with MNNG. With the help of complementation analysis, they could identify 14 complementation groups. Recombination between mutants of the same complementation group revealed their tight linkage (less than 0.05% wild type recombinants) which helped in construction of a tentative genetic map by the method of two factor crossing (Rimon and Oppenheim 1974). MNNG mutagenesis was also applied for isolation of *ts* clear mutants of LPP-2 SPI and was designated as LPP-2 SPI *cts* 1 (Rimon and Oppenheim 1975). This mutant formed clear plaques at 40^{0}C but turbid plaques at 26^{0}C on *P. boryanum.* The kinetics of protein synthesis as revealed by SDS-PAGE electrophoresis led to identification of 12 protein species that constitute the mature virus; and also 12 additional phage induced protein species. Presence of viral induced 'early' and 'late' proteins were also demonstrated.

(c) *Photoreactivation in hosts and cyanophages*

Mutagens affect DNA replication or repair. UV radiation (2537 A^0), one of the physical mutagens, produces intrastrand cyclobutane dimer between adjacent pyrimidines (Witkin 1976). Such errors are repaired either in light (photoreactivation) or in dark (dark/recombinational repair) by monomerization (splitting) or elimination of dimers, respectively (Singh 1967; Singh and Singh 1967b, 1972; Singh 1975). Photoreactivation occurs under light by photoreactivating enzyme (PRE).

Table 7. Spontaneous and induced mutations in cyanophages.

Cyanophage	Genetic marker	Spontaneous mutation frequency: Forward	Spontaneous mutation frequency: Reversion	Induction by	References
AS-1	$m^+ \rightarrow m$	$2\text{-}20\times10^{-4}$	-	acriflavine	Amla (1979a)
	$h^+ \rightarrow h$	$10^{-9}\text{-}10^{-8}$	-	UV, MNNG	Amla (1979a)
LPP-1	$r^+ \rightarrow r$	$0.1\text{-}1.3\times10^{-3}$	8.5×10^{-3}	UV, acriflavie	Singh *et al.* (1972)
	$h^+ \rightarrow h$	$5.2\text{-}17\times10^{-8}$	$0.53\text{-}15.6\times10^{0}$	UV, MNNG, 2-AP acriflavine	Singh & Kashyap (1976) Singh & Kashyap (1976) Kashyap (1980)
	$ts^+ \rightarrow ts$	$0.6\text{-}1.7\times10^{-3}$	1.3×10^{-4}	UV, MNNG, 2-AP acriflavine	Singh & Kashyap (1976)
LPP-2 SPI	$ts^+ \rightarrow ts$	-	-	MNNG	Rimon & Oppenheim (1974)

UV - ultraviolet; 2 AP - 2-aminopurine; MNNG - N'-methyl-N-nitro-N-nitrosoguanidine

m - minute plaque; *h* - host range; *r* - rapid lysis; *ts* - temperature sensitive

Studies on photorepair system of host as well as phage following UV exposure has been investigated in LPP-1 (Singh 1967; Singh and Singh 1967b, 1972; Wu *et al.* 1967). Both *Plectonema boryanum* and LPP-1 was observed to be photoreactivable in black, blue and white light (Singh 1967; Singh and Singh 1967b; Wu *et al.* 1967; Singh and Singh 1972; Singh 1975) and the optimum duration of light was about 6 h (Singh and Singh 1972). Mutagenic induction of rapid lysis mutation (*r*) induced by UV light was observed to be photorepaired by white light (Singh and Kashyap 1976). Cyanophage P-2 exhibited more or less similar system as LPP-1 (Singh and Singh 1972).

On the other hand, *Cylindrospermum* was found to be capable of photoreactivation in all black, blue, green, yellow, red and white light (Singh 1975). Photorepair in N-1 is strongly dependent upon the intensity of incident light. Blue and red light was effective as compared to black, green and yellow lights.

In AS-1 and its host *A. nidulans,* black, blue and white lights have been shown to affect photorecovery (Asato 1976; Amla 1979b). One of the partially photoreactivable mutant of *Anacystis nidulans* showed partial reactivation of virus in black light (Asato 1976).

(d) *Dark repair*

In blue-green algae the pyrimidine dimers are repaired in dark by excision from DNA by endonuclease II (Hanawalt 1975) followed by repair during replication. This mechanism was reported in *Anacystis nidulans* after observing the post-treatment effect of acridine orange, acriflavin (Singh 1967; Singh and Singh 1967c, 1972; Singh *et al.* 1969; Singh 1975) and caffeine (Asato 1972). In UV sensitive mutants of *Synechocystis aquatilis*, only dark repair system is observed (Zhevner and Shestakov 1972).

N-1 has its own dark rapair system of genome which is not interfered by host alga (Kashyap 1978). The mutagenic action of ultraviolet light is more pronounced in dark than in light which was investigated using the *r, h* and *ts* markers in cyanophage LPP-1 (Singh and Kashyap 1976).

REFERENCES

Adolph, K.W. and Haselkorn, R. (1971) Isolation and characterization of a virus infecting the blue-green algae *Nostoc muscorum. Virology* **46:** 200-208.

Adolph, K.W. and Haselkorn, R. (1972a) Photosynthesis and the development of blue-green algal virus N-1. *Virology* **47:** 370-374.

Adolph, K.W. and Haselkorn, R. (1972b) Comparison of the structures of blue-green algal LPP-1M and LPP-2 and bacteriophage T7. *Virology* **47:** 701-710.

Adolph, K.W. and Haselkorn, R. (1973a) Blue-green algal virus N-1: Physical properties and dissembly into structural parts. *Virology* **53:** 427-440.

Adolph, K.W. and Haselkorn, R. (1973b) Isolation and characterization of a virus infecting blue-green alga of the genus *Synechococcus. Virology* **54:** 230-236.

Allen, M.M. and Hutchinson, F. (1976) Effect of some environmental factors on cyanophage AS-1 development in *Anacystis nidulans. Arch. Microbiol.* **110:** 55-60.

Allen, M.M. and Smith, A.J. (1969) Nitrogen chlorosis in blue-green algae. *Arch. Microbiol.* **69:** 114-120.

Amla, D.V. (1979a) Mutagenesis of free and intracellular cyanophage AS-1 by UV, N-methyl-N-nitro-N-nitrosoguanidine and acriflavin. *Mutat. Res.* **59:** 147-156.

Amla, D.V. (1979b) Photoreactivation of ultraviolet irradiated blue-green alga *Anacystis nidulans* and cyanophage AS-1. *Arch. Virology* **59:** 173-179.

Amla, D.V., Rowell, P. and Stewart, W.D.P. (1987) Metabolic changes associated with cyanophage N-1 infection of the cyanobacterium *Nostoc muscorum. Arch. Microbiol.* **148:** 321-327.

Asato, Y. (1972) Isolation and characterization of ultraviolet sensitive mutants of the blue-green alga *Anacystis nidulans. J. Bacteriol.* **110:** 1058-1064.

Asato, Y. (1976) Ultraviolet light inactivation and photoreactivation of AS-1 cyanophage and *Anacystis nidulans. J. Bacteriol.* **126:** 550-552.

Bagchi, S.N., Kaloya, P. and Bisen, P.S. (1987) Effect of cyanophage N-1 infection on the synthesis and stability of *Nostoc muscorum* nitrate reductase. *Current Microbiol.* **15:** 61-65.

Balogh, A., Borbely, G., Cscke, C.S., Udvardy, J. and Farkas, G.L. (1979) Virus infection affects the molecular properties and activity of glucose-6-P-dehydrogenase in *Anacystis nidulans,* a cyanobacterium. *FEBS Lett.* **105:** 158-162.

Booth, S.J. (1975) Isolation, identification and partial characterization of cyanophage LPP-2N. *Ph.D. Thesis,* University of Nebraska, USA.

Bradley, D.E. (1967) Ultrastructure of bacteriophages and bacteriocins. *Bacteriol. Rev.* **31:** 230-314.

Brown, R.M. Jr. (1972) Algal viruses. *Advan. Virus Res.* **17:** 243-277.

Brown, R.M. Jr., Smith, K.M. and Walne, P.L. (1966) Replication cycle of the blue-green algal virus LPP-1. *Nature* **212:** 729-730.

Cannon, R.E. and Shane, M.S. (1972) The effect of antibiotic stress on protein synthesis in the establishment of lysogeny of *Plectonema boryanum. Virology* **49:** 130-133.

Cannon, R.E., Shane, M.S. and Bush, V.N. (1971) Lysogeny of a blue-green alga *Plectonema boryanum. Virology* **45:** 149-153.

Chao, L. and Bowen, C.C. (1971) Purification and properties of glycogen isolated from a blue-green alga *Nostoc muscorum. J. Bacteriol.* **105:** 331-338.

Daft, M.J., Begg, J. and Stewart, W.D.P. (1970) A virus for blue-green algae from fresh water habitats in Scotland. *New Phytol.* **69:** 1029-1038.

Edelman, M., Swinton, D., Schiff, J.A., Epstein, H.T. and Zeldin, B. (1967) Deoxyribonucleic acid of the blue-green algae (cyanophyta). *Bacteriol. Rev.* **31:** 315-331.

Fox, J.A., Booth, S.J. and Martin, E.L. (1976) Cyanophage SM-2: a new blue-green algal virus. *Virology* **73:** 557-560.

Franche, C. (1987) Isolation and characterization of a temperate cyanophage for a tropical *Anabaena* strain. *Arch. Microbiol.* **148:** 172-177.

Ginzburg, D., Padan, E. and Shilo, M. (1968) Effect of cyanophage infection on CO_2 photoassimilation in *Plectonema boryanum. Virology* **2:** 695-701.

Ginzburg, D., Padan, E. and Shilo, M. (1976) Metabolic aspects of LPP cyanophage replication in the cyanobacterium *Plectonema boryanum. Biochim. et Biophys. Acta* **423:** 440-449.

Goldstein, D.A. and Bendet, I.J. (1967) Physical properties of the DNA from the blue-green algal virus LPP-1. *Virology* **32:** 614-618.

Goldstein, D.A., Bendet, I.J., Lauffer, M.A. and Smith, K.M. (1967) Some biological and physico-chemical properties of blue-green algal virus LPP-1. *Virology* **32:** 601-613.

Goryushin, V.A. and Chaplinskaya, S.M. (1966) Existence of viruses of blue-green algae. *Mikrobiol. Zh. Akad. Mauk. Ukr. SRR.* **28:** 94-97.

Goryushin, V.A. and Chaplinskaya, S.M. (1968) Finding of the viruses lysing blue-green algae. In: *Blooming Waters,* pp. 171-174. Scientific Thought Publishing House, Kiev.

Granhall, U. and Hofsten, A.V. (1969) The ultrastructure of a cyanophage attack of *Anabaena variabilis. Physiologia Plant.* **22:** 713-722.

Gromov, B.V. (1983) Cyanophages. Ann. Microbiol. **134:** 43-59.

Gromov, B.V. and Kozyakov, S.Ya. (1970) A study of the peculiarities of the interrelationship between a blue-green algal population, *Plectonema boryanum* and cyanophage LPP-1. *Bull. Leningrad. Univ.* **3:** 128-135.

Gromov, B.V., Kozyakov, S.Ya., Mankaeva, K.A. and Gaevskaya, E.I. (1974) The ultrastructure of the cyanophage A-1(L) developments in the cells of blue-green alga *Anabaena variabilis. Bull. Acad. Sci. USSR Biological* **2:** 286-288.

Hanawalt, P. (1975) Molecular mechanisms involved in DNA repair. *Genetics* **79:** 179-197.

Haselkorn, R. (1978) Heterocysts. *Ann. Rev. Plant Physiol.* **29:** 319-344.

Hu, N., Thiel, T., Giddings, T.H. and Wolk, C.P. (1981) New *Anabaena* and *Nostoc* cyanophages from sewage settling ponds. *Virology* **114:** 236-246.

Jackson, D. and Sladecek, V. (1970) Algal viruses - eutrophication control. *Yale Scientific Magazine* **44:** 16-21.

Jensen, T.E. and Boven, C.C. (1970) Cytology of blue-green algae. II. Unusual inclusions in the cytoplasm. *Cytologia* **35:** 132-152.

Kashyap, A.K. (1978) Photoreactivation and dark repair of ultraviolet damage in the cyanophage N-1. *Advances in Cyanophyte Res.,* pp. 125-130.

Kashyap, A.K. (1980) Properties of spontaneous host range mutants of cyanophage LPP-1. *New Botanist* **7:** 23-29.

Kashyap, A.K. and Singh, Surendra (1989) Changes in photoelectron transport activity in cyanophage N-1 infected cells of *Nostoc muscorum. Current Microbiol.* **18:** 151-157.

Kashyap, A.K., Rai, A.N. and Singh, Surendra (1988) Effect of cyanophage N-1 development on nitrogen metabolism of the cyanobacterium *Nostoc muscorum. FEMS Microbiol. Lett.* **51:** 145-148.

Karbisheva, E.A., Goryushin, V.A., Mikhailyuk, D.P. and Shestakov, S.V. (1974) A study on the survival of cyanophage AM-1 irradiated with UV and X-rays in cells of radiosensitive mutants of the blue-green alga *Anacystis nidulans*. *Biol. Nauki* **17:** 118-121.

Kaushik, B.D. and Venkataraman, G.S. (1982) Cyanophages: A review. *Phykos* **21:** 64-72.

Khudiakov, I.Ya. (1977) Characterization of a new cyanophage S-2L lysing the unicellular cyanobacterium belonging to *Synechococcus* genus. *T. XLVI*: 547-552.

Khudiakov, I. Ya. and Gromov, B.V. (1973) The temperate cyanophage A-4(L) of the blue-green alga *Anabaena variabilis* strain (458). *Mikrobiologiya* **42:** 904-907.

Kirillova, F.M. and Chaplinskaya, S.M. (1973) Morphogenesis of the virus of blue-green algae studied by electron microscopy. *Mikrobiologiya* **42:** 570.

Kozyakov, S.Ya. (1974) A study of the development of cyanophage A-1(L) in a culture of the blue-green alga *Anabaena variabilis. Vestnik Leningradskogo Universitat* **15:** 102-108.

Kozyakov, S.Ya., Gromov, B.V. and Khudiakov, I.Ya. (1972) A-1(L) cyanophage of the blue-green alga *Anabaena variabilis. Mikrobiologiya* **49:** 555-559.

Lang, N.J. (1968) The fine structure of blue-green algae. *Ann. Rev. Microbiol.* **22:** 15-46.

Lark, K.G. and Adams, M.H. (1953) The stability of the phages as a function of the ionic environment. *Cold Spring Harbor Symp. Quant. Biol.* **18:** 171.

Lijas, A. and Rossman, M.G. (1974) X-ray studies of protein interaction. *Ann. Rev. Biochem.* **43:** 475-507.

Luftig, R. and Haselkorn, R. (1967) Morphology of a virus of blue-green algae and properties of its deoxyribonucleic acid. *J. Virology* **1:** 344-361.

Luftig, R. and Haselkorn, R. (1968a) Studies on the structure of blue-green algae virus LPP-1. *Virology* **34:** 664-667.

Luftig, R. and Haselkorn, R. (1968b) Comparison of blue-green algae virus LPP-1 and the morphologically related viruses G III and coliphage T_7. *Virology* **34:** 675-678.

Mackenzie, J.J. and Haselkorn, R. (1972a) Physical properties of blue-green algal virus SM-1 and its DNA. *Virology* **49:** 497-504.

Mackenzie, J.J. and Haselkorn, R. (1972b) An electron microscope study of infection by the blue-green algal virus SM-1. *Virology* **49:** 505-516.

Mendzhul, M.I. and Zhigir, V.V. (1971) Electron microscopic investigation of the caudal process structure of the virus LPP-1 isolate. *Microbiological J. (Kiev)* **33:** 460-464.

Mendzhul, M.I., Lisenko, T.G. and Bobrovnik, S.A. (1974) Effect of certain physicochemical factors on reproduction of viruses A-1 and LPP-1A in alga cells. Microbiological J. Kiev. **36:** 738-742.

Mendzhul, M.I., Lisenko, T.G. Bobrovnik, S.P., Sahun, T.S. and Nesterova, N.V. (1975) Detection of blue-green algae virus with octahedral type of symmetry. *Microbiological J.* **37:** 713-717.

Moskovets, S.M., Nesterova, N.V., Votselko, S.K., Stepanyuk, V.V., Mendzhul, M.I. and Pilipenko, U.G. (1971) *Mikrobiol. Zhur.* **33:** 583-589.

Padan, E. and Shilo, M. (1969) Distribution of cyanophage in natural habitats. *Verh. Internat. Verlin. Limnol.* **17:** 747-751.

Padan, E. and Shilo, M. (1973) Cyanophages - Viruses attacking blue-green algae. *Bacteriol. Rev.* **37:** 343-370.

Padan, E., Ginzburg, D. and Shilo, M. (1970) The reproductive cycle of cyanophage LPP-1G in *Plectonema boryanum* and its dependence on photosynthetic and respiratory systems. *Virology* **40:** 514-521.

Padan, E., Rimon, A., Ginzburg, D., and Shilo, M. (1971) A thermosensitive cyanophage (LPP-1-G) attacking the blue-green alga *Plectonema boryanum. Virology* **45:** 773-776.

Padan, E., Shilo, M. and Kislev, N. (1967) Isolation of 'cyanophages' from fresh water ponds and their interaction with *Plectonema boryanum. Virology* **32:** 234-246.

Padan, E., Shilo, M. and Oppenheim, A.B. (1972) Lysogeny of the blue-green alga by LPP-2-SPI cyanophage. *Virology* **47:** 525-526.

Padhy, R.N. and Singh, P.K. (1976) Mutation of resistance for virus N-1 in the blue-green alga *Nostoc muscorum. Arch. Virology* **52:** 85-90.

Padhy, R.N. and Singh, P.K. (1977a) Effect of physical and chemical agents on the blue-green algal virus N-1. *Acta Virologica* **21:** 264-267.

Padhy, R.N. and Singh, P.K. (1977b) Effect of temperatures on the adsorption and one step growth of the *Nostoc* virus N-1. *Arch. Mikrobiol.* **115:** 163-167.

Padhy, R.N. and Singh, P.K. (1977c) Effect of the blue-green algal virus N-1. *Mikrobios Letters* **5:** 135-139.

Padhy, R.N. and Singh, P.K. (1978a) Stabilizing effects of metallic ions in the blue-green algal virus N-1. *Biochem. Physiol. Pflanzen.* **173:** 188-192.

Padhy, R.N. and Singh, P.K. (1978b) Effects of host ageing, ions and pH on the adsorption of cyano virus N-1 to *Nostoc muscorum. Arch. Microbiol.* **116:** 289-292.

Padhy, R.N. and Singh, P.K. (1978c) Reversion of virus N-1 resistant mutant of blue-green alga *Nostoc muscorum. Experientia* **34:** 1565.

Padhy, R.N. and Singh, P.K. (1978d) Lysogeny in the blue-green alga *Nostoc muscorum. Arch. Microbiol.* **117:** 265-268.

Pearson, N.J., Small, E.A. and Allen, M.M. (1975) Electron microscopic study of the infection of *Anacystis nidulans* by cyanophage AS-1. *Virology* **65:** 469-479.

Rimon, A. and Oppenheim, A.B. (1974) Isolation and genetic mapping of temperature sensitive mutants of cyanophage LPP-2 SPI. *Virology* **62:** 981-986.

Rimon, A. and Oppenheim, A.B. (1975) Heat induction of the blue-green alga *Plectonema boryanum* lysogenic for cyanophage SPI *cts* 1. *Virology* **64:** 454-463.

Safferman, R.S. (1973) Phycoviruses. In: *The Biology of Blue-green Algae* (ed. N.G. Carr and B.A. Whitton), pp. 214-237. Univ. of California Press.

Safferman, R.S. and Morris, M.E. (1963) Algal virus: Isolation. Science **140:** 679-680.

Safferman, R.S. and Morris, M.E. (1964a) Growth characteristics of the blue-green algal virus LPP-1. J. Bacteriol. **88:** 771-775.

Safferman, R.S. and Morris, M.E. (1964b) Control of algae with viruses. *J. Amer. Water Wks. Assoc.* 56: 1217-1224.

Safferman, R.S. and Morris, M.E. (1967) Observations on the occurrence, distribution and seasonal incidence of blue-green algal viruses. *Appl. Microbiol.* **15:** 1219-1222.

Safferman, R.S., Diener, T.O., Desjardin, P.R. and Morris, M.E. (1972) Isolation and characterization of AS-1, a phycovirus infecting the blue-green algae *Anacystis nidulans* and *Synechococcus cedrorum. Virology* **47:** 105-113.

Safferman, R.S., Morris, M.E., Sherman, L.A. and Haselkorn, R. (1969a) Serological and electron microscopic characterization of a new group of blue-green algal viruses (LPP-2). *Virology* **39:** 775-780.

Safferman, R.S., Schneider, I.R., Steere, R.L., Morris, M.E. and Diener, T.O. (1969b) Phyco-virus SM-1: a virus infecting unicellular blue-green algae. *Virology* **37: 386-395.**

Samimi, B. and Drews, G. (1978) Adsorption of cyanophage AS-1 to unicellular cyanobacteria and isolation of receptor material from *Anacystis nidulans. J. Virology* **25: 164-174.**

Schneider, I.R., Diener, T.O. and Safferman, R.S. (1964) Blue-green algal virus LPP-1: purification and partial characterization. *Science* **114: 1127-1130.**

Shane, M.S. (1971) Distribution of blue-green algal viruses in various types of natural waters. *Water Res.* **5: 711-716.**

Sherman, L.A. (1970) The structure and replication of the blue-green algal virus LPP-1. *Ph.D. Thesis,* Univ. of Chicago, USA.

Sherman, L.A. (1976) Infection of *Synechococcus cedrorum* by the cyanophage AS-1M. III. Cellular metabolism and phage development. *Virology* **71: 199-206.**

Sherman, L.A. and Brown, R.M. Jr. (1978) Cyanophages and viruses of eukaryotic algae. In: *Comprehensive Virology* (eds. H. Frankel-Conrat and R.R. Wagner), pp. 145-234. Plenum Press, New York.

Sherman, L.A. and Connelly, M. (1976) Isolation and characterizataion of a cyanophage infecting the unicellular blue-green algae *A. nidulans* and *S. cedrorum. Virology* **72: 540-544.**

Sherman, L.A. and Haselkorn, R. (1970a) LPP-1 infection of the blue-green alga *Plectonema boryanum*. I. Electron microscopy. *J. Virology* **6: 820-833.**

Sherman, L.A. and Haselkorn, R. (1970b) LPP-1 infection of blue-green alga *Plectonema boryanum* II. Viral deoxyribonucleic acid synthesis and host deoxyribonucleic acid breakdown. *Virology* **6: 834-840.**

Sherman, L.A. and Haselkorn, R. (1970c) LPP-1 infection of the blue-green alga *Plectonema boryanum*. III. Protein synthesis. *Virology* **6; 841-846.**

Sherman, L.A. and Haselkorn, R. (1970) Growth of blue-green algal virus LPP-1 under conditions which impair photosynthesis. *Virology* **45: 739-746.**

Sherman, L.A. and Pauw, P. (1976) Infection of *Synechococcus cedrorum* by cyanophage AS-1M. II. Protein and DNA synthesis. *Virology* **71: 17-27.**

Sherman, L.A., Connelly, M. and Sherman, D.M. (1976) Infection of *Synechococcus cedrorum* by cyanophage AS-1M. I. Ultrastructure of infection and phage assembly. *Virology* **71**: 1-16.

Shilo, M. (1971) Biological agents which cause lysis of blue-green algae. *Mitt. Internat. Verein. Limnol.* **19**: 206-213.

Shilo, M. (1972) The ecology of cyanophages. *Bamidgeh* **24**: 76-82.

Silverberg, J., Rimon, A., Kessel, M. and Oppenheim, A.B. (1977) Assembly site of cyanophage LPP-2 SPI in *Plectonema boryanum*. *Virology* **77**: 437.

Simon, R.D. (1971) Cyanophycin granules from the blue-green alga *Anabaena cylindrica*: a reserve material consisting of copolymers of aspartic acid and arginine. *Proc. Nat. Acad. Sci. USA.* **68**: 265-267.

Singh, P.K. (1967) Ultraviolet sensitivity and genetics of blue-green algae and their viruses. *Ph.D. Thesis,* Banaras Hindu University, Varanasi.

Singh, P.K. (1973) Occurrence and distribution of cyanophages in ponds, sewage and rice fields. *Arch. Mikrobiol.* **89**: 169-172.

Singh, P.K. (1974) Isolation and characterization of a new virus infecting the blue-green alga *Plectonema boryanun*. *Virology* **58**: 586-588.

Singh, P.K. (1975) Photoreactivation of UV-irradiated blue-green algae and algal virus LPP-1. *Arch. Microbiol.* **103**: 297-302.

Singh, P.K. and Padhy, R.N. (1982) Blue-green algal viruses cyanophages. *Indian Rev. Life Sci.* **2**: 213-238.

Singh, R.N. and Kashyap, A.K. (1976) Mutagenesis in cyanophage LPP-1. *Mutat. Res.* **37**: 19-26.

Singh, R.N. and Kashyap, A.K. (1977) Induction of mutation in the blue-green alga *Plectonema boryanum* Gomont. *Mutat. Res.* **43**: 37-44.

Singh, R.N. and Singh, P.K. (1967a) Isolation of cyanophages from India. *Nature* **216**: 1020-1021.

Singh, R.N. and Singh, P.K. (1967b) Ultraviolet and photodynamic inactivation and photoreactivation of cyanophages and blue-green algae. *Proc. 54th Indian Sci. Congr.* Botany Section.

Singh, R.N. and Singh, P.K. (1967c) Pre - and post-treatment modification of ultraviolet lesions in blue-green algae with acriflavin. *Proc. 54th Indian Sci. Congr.* Botany Section.

Singh, R.N. and Singh, P.K. (1972) Ultraviolet damage modification and repair of blue-green algae and their viruses. In: *Taxonomy and Biology of Blue-green Algae* (ed. T.V. Desikachary), pp. 246-257. Univ. of Madras Press, India.

Singh, R.N., Singh, P.K., Kashyap, A.K., Sarma, T.A., Dhar, B., Chaube, I.J. and Choudhury, I.D. (1972) Isolation and characterization of new cyanophages and mutation of LPP-1 and host alga *Plectonema boryanum.* In : *Taxonomy and Biology of Blue-green Algae* (ed. T.V. Desikachary), pp. 585-591. Univ. of Madras Press, India.

Singh, H.N., Kumar, H.D. and Prakash, G. (1969) Post-irradiation modification of ultraviolet sensitivity of normal and nitrosofurazone resistant strains of the blue-green alga *Anacystis nidulans. Radiat. Bot.* **9:** 105-110.

Smith, K.M., Brown, R.M. Jr., Goldstein, D.A. and Walne, P.L. (1966a) Culture methods for the blue-green alga *Plectonema boryanum* and its virus with an electron microscope study of the virus infected cells. *Virology* **28:** 580-591.

Smith, K.M., Brown, R.M. Jr., Walne, P.L. and Goldstein, D.A. (1966b) Electron microscopy of the infection process of the blue-green algal virus. *Virology* **30:** 182-192.

Smith, K.M., Brown, R.M. Jr. and Walne, P.L. (1967) Ultrastructural and time lapse studies on the replication cycle of the blue-green algal virus LPP-1. *Virology* **31:** 329-337.

Stanier, R.Y. and Cohen-Bazire, G. (1977) Phototrophic prokaryotes: the cyanobacteria. *Ann. Rev. Microbiol.* **31:** 225-274.

Stewart, W.D.P. (1980) Some aspects of structure and function in N_2-fixing cyanobacteria. *Ann. Rev. Microbiol.* **34:** 497-536.

Stewart, W.D.P. and Daft, M.J. (1976) Algal lysing agents of fresh water habitats. In: *Microbiology in Agriculture, Fisheries and Food Symposium* Series 4 (eds. F.A. Skinner and J.G. Carr), pp. 63-90. Academic Press, New York.

Ueda, K. (1965) Virus like structures in the cells of the blue-green alga *Oscillatoria princeps. Exper. Cell Res.* **40:** 671-673.

Vallee, B.L. and Walker, W.C. (1970) *The protein* (ed. H. Neurath) 5: pp. 1-192. Academic Press, New York, 2nd edition.

Von Hippel, P.H. and McGhee, J.D. (1972) DNA-protein interactions. *Ann. Rev. Biochem.* **41:** 231-300.

Vorontsova, G.V., Karbysheva, E.A., Goryushin, V.A. and Shestakov, S.V. (1974) Effect of caffeine and acriflavin on survival of UV-irradiated cyanophage AM-1 in the cells of radiosensitive mutants of *Anacystis nidulans. Biol. Nauki.* **11:** 107-110.

Whitton, B.A., Carr, N.G. and Craig, I.W. (1971) A comparison of the fine structure and nucleic acid biochemistry of chloroplasts and blue-green algae. *Protoplasma* **72:** 325-357.

Witkin, E.M. (1976) Ultraviolet mutagenesis and inducible DNA repair in *Escherichia coli. Bacteriol. Rev.* **40:** 869-907.

Wu, J.H., Lewin, R.A. and Werkin, H. (1967) Photoreactivation of UV-irradiated blue-green algal virus LPP-1. *Virology* **31:** 657-664.

Wu, J.H. and Shugarman, P.M. (1967) Effect of virus infection on rate of photosynthesis and respiration of blue-green alga *Plectonema boryanum. Virology* **32:** 166-167.

Zhevner, V.D. and Shestakov, S.V. (1972) Studies on the ultraviolet sensitive mutants of blue-green alga *Synechocystis equatilis* Sanv. *Arch. Microbiol.* **86:** 349-360.

Ind. Phyco. Rev. 1993.
2: 113-138.

METHODS OF STUDY

B.D. KAUSHIK
National Facility for Blue-green Algal Collections,
Indian Agricultural Research Institute,
New Delhi - 110 012.

Blue-green algae are renowned for their ability to tolerate a wide range of environmental conditions, a characteristics of many primitive organisms. They probably represent, along with the photosynthetic and chemosynthetic bacteria, the survivors of the earliest photosynthetic plants. Their fossilized remains have been found in middle Precambrian rocks, and they may have appeared much earlier.

Their procaryotic cellular organization has restricted cell differentiation and confined them to microscopic size. Individual cells of plants or blue-greens are virtually invisible without magnification, and their identification usually requires the high power objective lens of a microscope to see adequately the morphological characteristics that separate genera, species and families.

I. FIELD OBSERVATION

1. *Soil algae*

Observation of an algal growth in its natural habitat are as important as its microscopic examination in the laboratory. Field study provides information about the site, environment, season and size and shape of the living algal thallus in its natural surroundings in the habitat. Many microalgae can be tentatively identified by their appearance, shape and colour of thallus in a particular habitat. For example, many species of *Oscillatoria* form a dark blue-green stratum at the bottom or on the surface of water of a continuously flowing fresh water channel. *Nostoc* thalli growing on dripping rocks looking like bunches of grapes, cannot be missed by an expert eye. Like *Oscillatoria*, *Cylindrospermum*, *Nostoc*, *Tolypothrix*, *Scytonema*, *Camptylonema*, *Aphanothece* and *Aphanocapsa*, form characteristic patches or strata on the surface of the soils of fallow lands and in crop fields. A large number of green algae, like network of *Hydrodictyon*, attached or floating filaments of *Oedogonium*, planktonic *Volvox* and masses of *Spirogyra* can be easily identified in the fields. Macroscopic thalli can be identified by their external appearance alone.

2. *Fresh water phytoplankton*

Free floating aquatic algae (the phytoplankton) may be collected from lakes, ponds, rivers and tanks by means of a plankton net. The relative proportions of different species represented in a plankton sample may be estimated by counting them on a haemacytometer. Fresh water phytoplankton may contain unicellular, colonial and simple filamentous algae, mostly belonging to Chlorophyceae, Cyanophyceae and Bacillariophyceae.

Freshwater phytoplankton, collected from relatively 'clean' nutrient deficient or oligotrophic waters, exhibit a great diversity of algal species, though their concentration as a whole may be very low. In contrast, polluted water bodies, which are rich in dissolved nutrients or are otherwise eutrophic, may at certain times, especially during the summer, have a luxuriant growth of one or rarely few species of algae, constituting what are known as "water blooms". The appearance of *Microcystis aeruginosa* bloom is regarded as indication of pollution. In such cases examination of a drop of water from water bodies containing bloom would reveal the presence of thousands of cells or colonies of the alga. The appearance of particular blooms often show a marked correlation with physical and chemical factors such as pH of water, light intensity and visibility, temperature, and nitrate, phosphate, oxygen or silica content. Methods of analysis of soil and water are beyond the scope of this chapter, however, the interested students may consult other books (Kaushik, 1987).

3. *Marine blue-green algae*

Although blue-greens are more abundant and in greater variety on land and in freshwater than in marine habitats, they are an important group among plants of the sea. There greatest abundance is in the intertidal zone, but they also occur in other zones.

Wherever there are rocks, seawalls or pilings, the upper intertidal zone is clearly marked by blue-greens. They form a "black zone" or upper intertidal band that is black or nearly so. In most habitats this band is almost exclusively *Calothrix crustacea*, but in some places, especially on creosoted pilings, the *Lyngbya confervoides* ecophene of *Microcoleus lyngbyaceus* is the colonizer.

The most luxuriant development of blue greens in the intertidal zone occurs in tropical marine waters. At the high tide line or even above the *Calothrix pilosa* form of *Scytonema hofmanii* forms a dense blackish turf on rocks and on the aerial roots of red and black mangroves. Somewhat lower down, but in the upper half of the intertidal zone, a variety of other species in many growth forms are observed.

Some of the same species that are predominant in the intertidal zone occur also below low tide to considerable depths but other species are found only in subtidal waters. In general, the blue-greens are not conspicuous in deeper waters. In a few habitats such as protected bays of muddy or muddy sand bottoms, they may form an extensive mat or patches. Even if they are not evident, they may be present in abubdance. Virtually al! larger algae have numerous blue-green epiphytes. These blue-green epiphytes are omnipresent to almost the greatest depths at which the host plant occur.

Blue-green algae have been reported many times in unexplained abundance from various depths below the compensation plane for photosynthetic organism, usually as plankton. Apparently they are living heterotrophically or chemiautosynthetically and are facultative photosynthetic.

Although a number of species of blue-green algae are often found as plankton, apparently only one species is strictly planktonic *i.e. Oscillatoria erythraea*. It is reported to occur in abundance in warmest months. *O. miniata* is an another planktonic species from *Microcoleus lyngbyaceus.*

The most assiduous penetrators of limestone of all algae are the blue-greens. Any shell or piece of limestone in the intertidal zone or below that has a greenish tinge is very likely to have a growth of blue-green filaments within to a depth of 1-5 mm, depending upon light penetration. The alga can be studied after a piece of the limestone is decalcified in dilute hydrochloric acid. The acid treatment does not seriously alter morphology of the plant.

Entophysalis of the family Chamaesiphonaceae produces strata or cushions or coccoid cells on the surface of a substratum. The filamentous endolithic growth form of this alga has been known under the genus, *Hyella,* whereas the coccoid cells on the surface have been placed in a number of genera including *Dermocarpa, Xenococcus, Pleurocapsa,* and *Aphanocapsa* as well as *Entophysalis.*

Blue-green algae that secrete organic acids and bore into limestone are important agents for the destruction of all forms of calcium carbonate in shallow waters. The important carbonate borers are *Mastigocoleus testarum, Plectonema terebrans, Calothrix* sp., *Hyella* sp. and *Scytonema* sp.

II. COLLECTION

Methods for collection of algal growths differ with the habit and habitat of an alga. Soil algae may be collected with the help of a sharp scalpel or knife. While collecting, the adhering soil should be avoided as far as possible. Algal samples collected from terrestrial habitats can be separated from soil particles by careful washing with distilled water. Soil algae should be collected in large size petridishes so that the growth is not disturbed and it should not be allowed to dry. Subaerial algae from moist rocks, walls, tree trunks and from other similar situations is also collected with a sharp knife and stored separately.

Planktonic algae are collected by using a plankton net of fine bolting silk of 173 meshes to the inch and the collected samples are stored in polypropylene bottles, large glass jars etc. Smaller forms (nannoplanktons) are collected by filtration of known volume of water through millipore membrane filter of 0.45 μ to 0.80 μ pore size. It is always desirable to add some quantity of water of the habitat so that the alga do not die for want of oxygen. Observable algal growths on the water surface or within water like on submerged twigs, weeds and stone can be collected using ordinary collection nets or with hands. For such collections, use of hand-lens or travelling microscope is desirable for tentative identifications of many algae and sorting them out of undesirable ones right in the field.

Collection of blue-green algae in marine environment often requires removal of some of the substrate. In the case of rocks etc. a hammer is a handy tool. On intertidal sand beaches protected from heavy wave action, a sample of sand should be collected to a depth of approximately 2 cm. In the laboratory, sample should be placed in a screw capped vial and shaken with several volumes of filtered seawater. Immediately after the sand has settled, the supernatant should be poured off into another container. If the concentration is too low it can be concentrated by centrifugation.

Ordinarily a collection bottle should not be filled with any algal sample more than a quarter of its capacity and the remainder part of the bottle be filled with water of the habitat. Large quantity of algal thalli may deplete the oxygen and deteriorates the material for preservation.

Fresh algal collection should be examined for studying the structure of thallus in living conditions. Oscillating forms can be best examined in living state. Therefore, collected algae should be maintained in the natural substratum (soil, water etc.) until they are killed. The killing and hardening (fixing) reagents must be used as soon as possible after proper collection.

III. ABUNDANCE

1. *Soil algae*

The relative growth or abundance of algae is estimated by means of microquadrats or microtransects. The microquadrats are laid down by placing 20 cm square frame in the field and drawing a corresponding outline on a graph paper. The contours and shapes of algal strata caused by different species are drawn to scale on the graph paper.

If the topography of the habitat is uneven, a microtransect may yield more reliable information. It is plotted by laying a string and then locating the algal patches in quadrats laid at fixed points along the transect. The actual contour of the habitat along the string and the location of different patches are then plotted to an appropriate scale.

Depending on the size of the stratum formed, blocks of soil of a certain fixed size (2, 5 or 10 mm^3) can be cut by means of sharp blade and the algal growth estimated by extracting the pigments from the block in a fixed volume of 80% acetone. Pigments should be drawn out in atleast three changes of acetone to ensure complete extraction. The measurement of absorbance of the pigment extract in a cuvette of fixed path length in a spectrophotometer at 490 mm (corresponding to myxoxanthin), against 80% acetone as blank. It will provide a rough estimate of the blue-green algae present in the original block. The algal growth should be correlated with variations in soil characteristics by regular soil analysis like moisture and organic content, pH, nitrate and carbonate content etc.

2. *Phytoplankton*

1. A known volume (250 ml) of formalin preserved phytoplankton is allowed to stand undisturbed in a measuring cylinder for 24 hr. The supernatant is siphoned off.

2. The sediment is swirled and homogenous suspension is poured into a 25 ml volumetric flask. Wash the sedimentation cylinder atleast twice and add washings into flask. Make up the volume and mix thoroughly.

3. Place a whipple eye piece micrometer into the ocular lens of a microscope. This micrometer is divided into 100 squares, with each square is further subdivided into 25 smaller squares. The area of each square can be determined by calibrating against a stage micrometer under low power microscope.

4. Fill the sedgwick-Rafter counting chamber with 1 ml of plankton concentrate, care is to be taken to avoid the air bubbles. Place specific coverslip and count after 10 min.

5. Determine the number of individuals of each species in 1 ml volume. The field count volume can be calculated by multiplying the field area by 1 mm (depth of chamber) by the number of squares counted. Finally multiply by the concentration factor.

For smaller forms (nannoplankton) a Palmer - Nannoplankton counting chamber can be employed. The Palmer chamber is 0.4 mm deep and is designed for counting 0.1 ml volumes. It is to be calibrated under high power of microscope (x40 or x45).

3. *Primary productivity*

1. Take known volume (20 ml) of aliquot from algae containing water bodies and centrifuge at 3000 g for 15 min. Discard the supernatant.

2. Suspend the pellet in 2-3 ml of distilled water and transfer quantitatively in a weighed crucible.

3. Dry the contents in an oven at 105^0C for 6 hr, cool in a desiccator and weigh. Calculate the algal biomass.

4. Place the weighed dry algae containing crucible in muffle furnace at 600^0C for 4 hr. Cool in desiccator and reweigh. The difference between algal dry weight and ash weight providing the biomass of the organism present.

$$\text{Daily production (gm)} = \text{Volume of water body } (m^3) \times \frac{\text{production in sample (gm)}}{\text{sample volume } (m^3)}$$

IV. KILLING AND PRESERVATION

Collected algae are either killed, preserved and stored for morphological studies ir the laboratory or as such fresh materials are carried for preparing museum or/and herbarium (for macroscopic forms) specimens. Material intended for culturing should be transported to laboratory without adding any preservatives.

When it is intended to store algae in the laboratory for subsequent morphological studies, they may be killed and preserved in a 4% solution of formalin (prepared by adding 4 ml of 40% formalin in 96 ml of distilled water; 40% commercial formalin is regarded as 100% for purpose of calculation). For preserving aquatic algae, an appropriate quantity of 40% formalin may be added directly to the sample so as to obtain a final concentration of 4%. The next most common killing and preservative reagent is FAA (Formalin, acetic acid and ethyl alcohol) and the composition is as follows:

	volume in ml.
Ethyl alcohol (95%)	50
Glacial acetic acid	5
Formaldehyde (37-40%)	10
Water	35

There are several modifications of FAA fixative and one such modification includes ethyl alcohol (50 or 70%), Glacial acetic acid and formalin in a ratio of 90:5:5. The following are the various fixatives and preservatives which can be employed to various groups of algae.

1. General fixative

a)	Water	93	ml
	Formaldehyde	5	ml
	Glacial acetic acid	2	ml
b)	Water	72	ml
	Formaldehyde	5	ml
	Glacial acetic acid	3	ml
	Glycerine	20	ml

2. For maintaining natural colour of algae the following two modifications of FAA have been also recommeded. The algae are immersed in the preservative for a few days and then transferred to FAA solution.

a)	Water	35	ml
	Copper sulphate*	2	gm
	Glacial acetic acid	5	ml
	Formaldehyde	10	ml
	Ethyl alcohol (95%)	50	ml

*Copper sulphate should be completely dissolved before adding acetic acid.

b)	Ethyl alcohol (50%)	90	ml
	Formaldehyde	5	ml
	Glycerine	2.5	ml
	Glacial acetic acid	2.5	ml
	Copper sulphate	10.0	gm
	Uranium nitrate	1.5	gm

Most material take 48 hr or more for fixation of their colour.

In addition to these, a large variety of fixative/preservative has been given which can be employed with specific objectives.

Marine algae are collected and preserved in 5 to 10% formalin in sea water. After about a week in formalin sea water, the thalli may be washed in graduated mixture of sea water and tap water and finally stored in 3-5% formalin in tap water with 5% glycerine and sufficient borax to prevent the solution from becoming acidic.

V. FIXATIVES AND PRESERVATIVES

Form-alcohol (Chicago formula)

Alcohol 70 percent	100	ml
Formalin	6	ml

Material may be fixed and stored in this fluid indefinitely.

Form-alcohol (Lynds Jones formula)

Alcohol 70 percent	100	ml
Formalin	2	ml

May be used either as a fixative or for storing material.

General fixative (King's formula)

Formalin 5 percent	100	ml
Alcohol 95 percent	150	ml
Glycerine	50	ml

Universal fixative (Rawlin's formula)

Alcohol 50 percent	100	ml
Formalin	6.5	ml
Glacial acetic acid	2.5	ml

Acetic alcohol fixative

Glacial acetic acid	1	part
Absolute alcohol	3	parts

Wash well in strong alcohol for 15 minutes after fixation and store material in 70 percent alcohol.

Carnoy's fixative

Absolute alcohol	2	parts
Chloroform	3	parts
Glacial acetic acid	1	part

Fix for 10-15 minutes and wash in 85 percent alcohol, in which the material may be stored.

Osmic acid fixative

Osmic acid	1	gm
Distilled water	100	ml

The solution must be kept in the dark. It is best prepared by breaking the clean tube in which it is bought, in the necessary distilled water. The water may be placed in the bottle, and the tube slipped in and broken by shaking. If exposed to the light metallic osmium is precipitated and the solution becomes useless. This may be avoided by dissolving in 1 percent chromic acid instead of distilled water.

Chromic acid fixative

Chromic acid	10	gm
Glacial acetic acid	10	gm
Water	1000	ml

Fix for 24 hours and wash in water.

Various other strengths of this fixative may be made up according to the nature of the material. The above is suitable for most cytological material; but a weaker solution, say, chromic acid 3 gm, glacial acetic acid 7 gm, water 1000 ml may sometimes be used.

Chromo-acetic fixative

Strong

Chromic acid 1 percent	140	ml
Glacial acetic acid	1	ml

Weak

Chromic acid 1 percent	140	ml
Glacial acetic acid	1	ml
Water	60	ml

Schaffner's chromo-acetic fixative

Chromic acid	0.3	gm
Glacial acetic acid	0.7	ml
Water	99.0	ml

A useful modification of the standard formula.

Benda's fixative

Chromic acid 1 percent	16	ml
Osmic acid 2 percent	4	ml
Glacial acetic acid	2	drops

Wash in water and grade through alcohol as for normal chromoacetic fixatives.

Licent's fixative

Chromic acid 1 percent	80	ml
Glacial acetic acid	5	ml
Formalin	15	ml

Chicago fixative

Chromic acid	1	gm
Glacial acetic acid	2	ml
Osmic acid 1 percent	6-8	ml
Water	90	ml

A very useful general fixative.

Chromo-acetic fixative for marine algae

Chromic acid	1.0	gm
Glacial acetic acid	0.4	ml
Sea water	400.0	ml

Material must be washed in sea water.

Navaschin's fixative

Solution A

Chromic acid	1.5	gm
Glacial acetic acid	10.0	ml
Distilled water	90.0	ml

Solution B

Formalin	40.0	ml
Distilled water	60.0	ml

Mix equal volume of solutions A and B immediately before use. A good fixative for mitotic figures in ferns and higher plants.

Rabl's fluid

Chromic acid 10 percent	7	ml
Formic acid (sp. gr. 1.2)	5	drops
Water	200	ml

May be used for fixing nuclei and mitotic figures.

Flemming's fixative

Strong

Chromic acid 1 percent	75	ml
Glacial acetic acid	5	ml
Osmic acid 2 percent	20	ml

Weak

Chromic acid 1 percent	25	ml
Acetic acid 1 percent	10	ml
Osmic acid 1 percent	10	ml
Water	50	ml

Flemming's solution (Newton's modification)

Chromic acid 1 percent	14	ml
Osmic acid 2 percent	4	ml
Glacial acetic acid	0.5	ml
Water	12	ml

Fix for about 6 hours. Wash for 24 hours in water.

N.K.L. Fixative (Navaschin-Karpechenko-Langlet formula)

Solution A

Chromic acid	1	gm
Acetic acid	10	ml
Water	65	ml

Solution B

Formalin	40	ml
Water	35	ml

Use in equal parts and keep cold. It will frequently be found better to use at half strength for a general fixative. Material may be left in the fixative for some time.

LA cour fixative

2BD	Chromic acid 1 percent	100	ml
	Pot. dichromate 1 percent	100	ml
	Acetic acid 5 percent	30	ml
	Osmic acid 2 percent	30	ml
	Saponin	0.1	gm
2BE	Chromic acid 1 percent	90	ml
	Acetic acid 5 percent	10	ml
	Osmic acid 2 percent	15	ml
	Pot. dichromate	1	gm
	Saponin	0.05	gm

Uranium trioxide may replace the osmic oxide. Bleach with hydrogen peroxide (1:4 of 80% alcohol). Stain with iodine-gentianviolet.

Smith's fixatives for pollen mother cells

S1	Chromic acid 1 percent	110	ml
	Osmic acid 2 percent	35	ml
	Acetic acid 5 percent	25	ml
	Pot. dichromate	0.5	gm
	Saponin	0.05	gm
	Distilled water	50	ml

S2	Chromic acid 1 percent	75	ml
	Osmic acid 2 percent	25	ml
	Acetic acid 5 percent	12.5	ml
	Pot. dichromate	1	gm
	Distilled water	46	ml

Gilson's fixative

Alcohol 95 percent	42.0	ml
Glacial acetic acid	18.0	ml
Nitric acid	2.0	ml
Mercuric chloride sat.aq.	11.0	ml
Water	60.0	ml

Wash with 70% alcohol after fixing for about 10 minutes.

Dichromate fixative (Champy formula)

Pot. dichromate 3 percent	7	parts
Chromic acid 1 percent	7	parts
Osmic acid 2 percent	1	part

Fix for 6-24 hours. Wash in running water for about the same time. The fixative is good for cytoplasmic structures.

Bensley's fixative

Mercuric chloride 2.5 percent	4	parts
Osmic acid 2 percent	1	parts

This fixative is used for mitochondria.

Mercury-alcohol fixative

Mercuric chloride	10	gm
Nitric acid	5	ml
Glacial acetic acid	2	ml
Alcohol 95 percent	30	ml
Water	270	ml

Fix for 1 hour and wash in 50 percent alcohol. Store in 70% alcohol.

Mercury-picric fixative

Mercuric chloride	5	gm
Glacial acetic acid	5	ml
Picric acid (sat.sol. in 30% alcohol)	100	ml

Fix for 1 hour and wash in 50% alcohol. Store in 70% alcohol.

Mercury-acetic fixative

Mercuric chloride	3	gm
Glacial acetic acid	3	ml
Alcohol 95 percent	100	ml

Fix for 1 hour and wash in 95% alcohol. Store material in 70% alcohol.

Zenker's fixative

Pot. dichromate	2.5	gm
Mercuric chloride	5.0	gm
Glacial acetic acid	5.0	gm
Water	100.0	ml

When mixed add 10 ml of formalin. Fix for 24 hrs and treat sections with iodine before staining.

Picric acid fixative

Picric acid	1	gm
Water or alcohol 70%	100	ml

This may be used cold for 1-24 hrs, or hot (85^{0}C) for 5-10 minutes. Washing must be continued until the yellow colour of the picric acid disappears.

Bouin's fluid

Picric acid (sat. aq. sol.)	75	ml
Formalin	25	ml
Glacial acetic acid	5	ml

This fixative should be allowed to act for 24 hrs. It may be left longer. Wash in water and grade up through alcohol. Store in 70% alcohol.

Bouin's fluid (Allen's modification)

Picric acid (sat. aq. sol.)	75	ml
Formalin	25	ml
Glacial acetic acid	5	ml

Add at time of fixing

Urea	2	gm
Chromic acid	1.5	gm

Heat fixative to 38^{0}C and fix for 1 hour. Cool to room temperature and dehydrate by dropping one drop of absolute alcohol per second into the fixative so that material has reached 75% alcohol in one hour. Wash out the picric acid with 75% alcohol containing a few drops of lithium carbonate. Shake frequently until the yellow colour has disappeared.

Picro-sulphuric fixative (Kleinenberg)

Picric acid (sat. aq. sol.)	100	ml
Sulphuric acid	2	ml
Distilled water	300	ml

This fixative is more generally useful for zoological than for botanical work, but may be found useful for anatomical work. The solution should be filtered before use.

Fixative for protoplasmic connections (Meyer)

Sulphuric acid	2	ml
Water	100	ml

Picric acid to sat. about 0.25 percent.

Fix for 2 hours, then wash out in 70% alcohol. Place fixed material in sulphuric acid of strength between 2 and 50% for at least half an hour before staining. Some material for example, moss leaves needs up to 24 hours in acid of 50% strength or over.

Merkel's fluid

Chromic acid 1 percent	25	ml
Platinic chloride 1 percent	25	ml
Water	150	ml

Recommended for nuclear division in higher plants. It is probably no better than chromo-acetic fixatives.

Hermann's fixative

Platinic chloride 1 percent	15	parts
Glacial acetic acid	1	part
Osmic acid 2 percent	2-4	parts

This is a modification of Flemming's solution and it is used for similar purposes and in a similar manner.

Iodine fixative

A saturated solution of iodine in a saturated solution of potassium iodide in water will be found a useful fixative for unicellular and colonial algae. Fix for 12-24 hrs wash in water, or, if starch is present, use a 0.5% solution of tannic acid in water.

Oxyquinoline sulphate (Chinosol) as preservative

Plant material may be easily and completely preserved in the field by packing in containers, preferably of glass or earthen ware, covering with water and adding enough chinosol to make a solution of strength up to 1 percent. The chinosol is in powder form, and the procedure has great advantages over formalin or alcohol, especially for travellers.

VI. PREPARATION OF HERBARIUM SPECIMENS

Cyanophycean members are not preserved as herbarium due to its size, however, for other larger forms herbarium specimens can be prepared.

The collected specimens from the sea water are directly placed on blotting paper/old newspaper, properly spread in shade in a room well ventilated, preferably not in humid room and allow to dry for some times. The alga must not be washed in tap water as the salt left in sea water keeps them flexible. For preparing dry herbarium specimens, these dried thalli should be soaked in sea water and then transferred to a solution of equal parts of water, ethyl alcohol, and glycerine with 10% phenol. The solution is allowed to penetrate thoroughly in tissues of thalli and then they are allowed to dry out on the herbarium sheet.

Large specimens are glued to the paper with water proof glue after they are dried completely.

Alternatively, specimens are floated in a flat tray half filled with water of the habitat. An adequate size of paper is placed beneath the specimen in the tray. The specimen is raised from one end out of water and spread properly with needle and brush. The excess water is slowly drained off, the paper along with specimen is lifted, placed on a drying blotter and covered with old linen sheet. The process is continued for 2 or more hours, for drying thoroughly before taking herbarium in stock.

VII. PREPARATION OF MOUNTS

Temporary mount

The commonest method of microscopic examination of algae is by mounting a little of the material on a slide in a drop of water. Such mounts are purely temporary as water will begin to dry after sometime and air cavities will appear between the slide and coverglass. When the material is to be studied for the extended period of time, the material is mounted in 10% glycerine with a drop of formalin and sealed with molten wax or quick drying paints. Sealing reduces the drying of mounting medium. Such mounts can be kept for longer period of time.

The following mounting medium may be used if the mount is desired to be kept for a longer period.

Phenol (molten)	20	ml
Lactic acid	20	ml
Glycerine	40	ml
Water	20	ml

If the material is desired to be stained, add 1% aqueous solution of cotton blue or aniline blue as follows:

Lactophenol	100	ml
Glacial acetic acid	<=20	ml
Dye solution	1-5	ml

Semi permanent mount

Schedule I

i. Place the alga on a slide in a drop of fresh or marine water.

ii. Add a small drop of 40% formalin to fix the algal material. Drain away the excess water.

iii. Place a small lump of glycerine jelly on the alga. (Glycerine jelly is prepared by dissolving 5 g gelatine in 30 ml of water by gentle heat at 35^{0}C and then adding 0.125 gm of phenol and 35 ml glycerol).

iv. Transfer the slide in an oven at 60^{0}C for a few min for the jelly to melt. Spread the algal material appropriately.

v. Place a coverglass and keep the slide once again in oven for a short time. Wipe off the excess jelly and seal the margins of coverglass with quick drying lacq or with Gold size.

Permanent mounts

For the preparation of permanently stained mounts weak or medium strength chromo-acetic acid is employed for filamentous alge.

Weak

10% aqueous chromic acid	2.5	ml
10% aqueous acetic acid	5	ml
Distilled water	92.5	ml

Medium

10% aqueous chromic acid	7	ml
10% aqueous acetic acid	10	ml
Distilled water	83	ml

Unicellular algae are better fixed in weak chromacetic acid solution while larger forms in medium. Filamentous marine algae may be fixed in 1% chromacetic acid fluid or in 10% formalin solution.

1% chromacetic acid

Chromic acid	1	gm
Glacial acetic acid	1	ml
Distilled water	99	ml

Schedule II

The alga should be washed thoroughly in running water while marine algae in sea water. It should be followed by washing the marine algae in sea-tap water and gradually increasing tap water ratio and finally in pure tap water followed by distilled water. Materials fixed in formalin-acetoalcohol should be washed in distilled water.

Washed material is placed in deep watch glass or dishes, excess liquid if any is removed with fine pipette and stained with suitable stains (Harris or Fe-Haematoxylin). The dye is differentiated with acid water and the differentiator is washed thoroughly with water. Stained material is dehydrated sequentially in 3, 5, 8, 12, 30, 50 and 70% alcohol for 20 min. in each. Then material is allowed to soak in 85% alcohol for atleast 18 hr, followed by 90 and 95% alcohol. In the end two changes are given in absolute alcohol. The material may be counter-stained if needed. Finally, the material is mounted in a drop of Euparal or Canada Balsum and covered.

Stains

1. *Harris Haematoxylin:* Dissolve 5 gm haematoxylin and 3 gm of ammonium alum (aluminium ammonium sulphate) in 1,000 ml 50% ethyl alcohol by heating. Add 6 gm mercuric oxide and boil for about half an hour, cooled, filtered and made to volume in a volumetric flask. Before making final volume 2-3 drops of HCl is added.

2. *Aniline blue:* Dissolve 1 gm in 100 ml of 90% alcohol. The algal material can be stained at 85% alcohol stage for 5 min.

3. *Erythrosine Bluish:* Dissolve 1 gm in 100 ml absolute alcohol. The alga can be stained after dehydrating in 95% ethyl alcohol by dipping for 30 sec.

4. *Light green:* Dissolve 0.2 gm of the stain in 100 ml of 90% alcohol. The algal material is stained for 30-60 sec. to stain cellulosic cell wall.

5. *Congo red:* Dissolve 0.2 gm of stain in 100 ml of alcohol. The cyanophycean mucilagenous sheath is stained by dipping in the solution after 95% alcohol stage for 1 min.

Schedule III

1. Smear the slide with a thin film of Meyer's albumin and allow it to dry for a few minutes. Place the material on this in a drop of water and tilt the slide to drain away the water. Pass the slide momentarily over a spirit flame.

2. Holding the slide in the left hand, pour 30% ethyl alcohol over it from a dropping bottle and immediately drain it away. Rinse the material in a similar manner with 50%, 70% 90% ethyl alcohol, using two changes in each grade. After the last change, immediately place a drop of Euparal on the material and apply a coverglass. If necessary, the material may be stained at the appropriate stage.

Schedule IV

This method is employed for making permanent slides suitable for cytological studies.

1. Fix the material in the fixatives described in Schedule II or in a mixture containing equal volumes of glacial acetic acid and absolute ethyl alcohol. The material should be immersed in the fixative for 10-15 min.

2. Wash the algal material on a slide in three changes of water.

3. Mordant in 1.5% aqueous iron alum solution for 10-60 sec.

4. Wash the slide thoroughly in running tap water for 2-3 min.

5. Stain the material in a drop or two of acetocarmine on a slide or, in the case of unicellular forms, in a centrifuge tube; warm the slide gently over a spirit flame and apply a coverglass.

6. Dip the slide and coverglass horizontally in a petri dish containing 95% ethyl alcohol and allow them to remain there till the coverglass separates from the slide. Carry the coverglass and slide through two other petri dishes containing absolute ethyl alcohol, mount in Euparal and apply the coverglass again.

For further details of this method, and its modifications for application to different kinds of algae, see Godward (1966).

Schedule V

This method is suitable for making fluid mounts of planktonic algae and has been suggested by Nygaard (1977).

1. Place a rectangular frame of soft paraffin, slightly smaller than the coverslip, on the slide. This can be conveniently dispensed from the tube containing paraffin if the screw cap is fitted with a tiny metallic capillary tube.

2. Place a drop of fixed phytoplankton in 4% formalin inside the paraffin frame (the fixative may be Lugol's solution or formalin).

3. Apply the coverslip over the drop and press gently, taking care to avoid air bubbles.

4. Seal the mount with an alcoholic solution of shellac and Venetian turpentine.

VIII. HISTOCHEMICAL AND GENERAL METHODS

Temporary fixation

Add a drop of dilute Iodine Potassium Iodide (IKI) solution to a 1-2 ml sample of algae; this acts also as a temporary preservative. (The IKI solution is prepared by dissolving 2 gm of KI in 20-30 ml of water and then dissolving 1 gm of iodine. The solution is then made up to 100 ml with distilled water).

Algae can be fixed and temporarily preserved also by exposure to osmic acid vapour.

General Staining

Different algae vary in their affinity for stains. More frequently employed stains are Methylene Blue, Gentian Violet and Acid Fuchsin (up to 1% aqueous solutions). The following simple procedure is recommended.

Mount algae in a drop of water on a slide and apply a coverglass. Add a drop of the stain to one edge of the coverglass and let it diffuse to the opposite edge by absorbing the water from the other side with a piece of dry blotting paper. In this way, a spectrum of staining is obtained, the algae near one side of the coverglass are intensely stained and on the opposite side weakly stained.

Arresting the Movements

In the living state, certain algae are actively motile or exhibit gliding movements. Others may exhibit a passive Brownian movement. To slow down such movements:

i) add a tiny drop of IKI solution to a drop of algal suspension;

ii) add a pinch of powdered gum arabic or dextran to a drop of algal suspension and apply a coverglass;

iii) mix a drop of chloroform water (i.e. a drop of chloroform in 5 ml of distilled water) with a drop of algal suspension and apply a coverglass; and

iv) place a drop of algal suspension directly in the centre of a slide or petri dish containing a thin sheet of 2.5% aqueous agar, apply a coverglass and observe under a microscope. This method will also arrest the Brownian movement.

Cell walls

The presence or absence of a true cell wall may be ascertained by microscopic observation of the algal cells that had been placed in a plasmolysing solution, e.g., 10% sucrose, for a few minutes.

The presence of cellulose or related compounds in a cell wall may be determined by treating the algae with Schultz's reagent (chlor-zinc-iodine) for 15-20 min., which gives a violet or blue colour to such compounds. Schultz's reagent used to be regarded as a specific stain for cellulose but it gives a positive colour response with other related compounds as well.

Alternatively, mix a drop of dilute IKI solution with a drop of algal suspension on a slide, and follow this either by treating the mixture with a drop of concentrated sulphuric acid or by irrigating with 70% sulphuric acid. Cellulose turns blue.

Cell walls composed largely of pectic substances acquire a red colour when stained with a fresh, dilute aqueous solution of Ruthenium Red (one or two crystals in a small watchglass containing 2-3 ml water).

Plasma Membranes

Thin sections or whole cells of many eucaryotic algae, when treated with 1% phosphotungstic acid in a 10% chromic acid solution (pH 3) for 15-20 min, often reveal well-stained plasma membranes (Sundberg and Lembi, 1976).

Gelatinous Envelopes and Sheaths

Gelatinous envelopes and sheaths are rendered conspicuous by staining with a very dilute aqueous solution of either Ruthenium Red or Methylene Blue. Alternatively, a trace of India ink or of Gurr's Negative Stain may be employed for the same purpose.

Chromatophores

Algal chromatophores are rendered conspicuous by treating the cells with boiling 8% aqueous silver nitrate solution for 4-5 min; this treatment turns the chromatophores brownish-black.

According to Friedmann (1966), the chromatophores assume a conspicuous and prominent appearance when viewed under a microscope which has a blue filter inserted between its light path. In visual work, this filter has been reported to produce a surprisingly clear image and in photomicrography its use yield sharp and brilliant negatives. A further advantage of the blue filter is that it can be used equally well with different kinds of algae since most of them have a strong yellow component (i.e., the complementary colour to blue) in their chromatophore pigments.

Nuclei and Chromosomes

Nuclei and chromosomes can be demonstrated by the application of the iron alum-acetocarmine technique, as described in Schedule IV.

In temporary preparations, nuclei may sometimes acquire a bluish colour when the cells are immersed in a very dilute aqueous solution of Methylene Blue, and then washed in water to remove the stain from the cytoplasm.

Starch and Other Reserve Materials

If a blue or violet colour develops after treatment of the material with the IKI solution, starch is indicated. Iodine in chloral hydrate gives better results in this test.

Glycogen or cyanophycean starch generally gives a reddish brown or maroon colour when treated with iodine whereas floridean starch gives a reddish colour with this reagent.

Sulphated and non-sulphated polysaccharides in algal tissues may be detected by treatment with alcian dyes. Parker and Diboll (1966) have recommended the following schedule for localization of acid and sulphated polysaccharides in various marine and freshwater algae;

i) fix the algae in FAA or in 10% formalin in seawater or freshwater, dehydrate in tertiary butyl alcohol series, embbed in paraffin under vacuum, cut sections (15 u thick) and mount on slides,

ii) stain 30-60 min in 0.5% aqueous Alcian Blue adjusted to pH 0.5 with N HCl,

iii) wash in water, and

iv) dehydrate, clear and mount.

Polyphosphate granules

Another nutrient reserve of blue-green algae, referred to in the past as volutin or metachromatic granules, occurs as acid soluble polyphosphates. They usually form in the centroplasm.

Cyanophycin granules

A type of granule that apparently found only in blue-green algae is known as cyanophycin. These granules are visible with light microscope without staining, at cross walls in all blue-greens except *Spirulina.* They tend to disappear when cultures are kept in the dark, are soluble in mineral acids, are broken down by enzyme pepsin, stained by acetocarmine, blackened by Osmium tetraoxide, and give a positive test for argenine.

Isolation of algae

Select a small quantity of algal thallus from homogenous looking spot. Examine a portion of it under the microscope, and, if found reasonably unialgal and free from fungi and actinomycetes, tease out the clump with the help of a pair of platinum needle under a stereoscopic binocular disecting microscope. Transfer the teased material to a few drops of sterile water to wash and pick up the one or two filaments with sterile platinum/nichrome inoculation needle and inoculate in a tube containing suitable growth medium. Alternatively, algal clumps are taken in a rubber or glass stoppered flask containing sterile media and glass beads. The contents are shaken vigorously for 10-15 min. The fragments of filaments/thallus are picked up by a fine capillary pipette and transferred to sterile watch glass with sterile medium. The algal material is washed 3-4 times and finally inoculated to agar medium in tubes. The inoculated tubes are incubated in light at a suitable temperature. After suitable incubation (10-15 days) algal growth is examined and if found contaminated, is purified further.

Silica gels (Pringsheim, 1914) have been found helpful, since a small proportion of single filaments transferred from such cultures to agar media often prove to be free from bacteria. Fogg (1942) obtained bacteria free cultures of *Anabaena cylindrica* by treating the alga with chlorine water (25 ppm) for 2 min. After washing with sterile water, it was plated and found devoid of contaminating organisms. Another promising method is to destroy the bacteria with the help of ultraviolet rays from a mercury vapour lamp. This was successfully achieved with *Anabaena variabilis* by Allison and Morris (1929). Bortels (1940), who adopted the same method, described it in detail. A small quantity of alga is dispersed in 2 ml of sterile water and exposed for 1 to 8 min in quartz tube to a quartz mercury vapour lamp, placed at a distance of 20 cm, shaking appears to be essential, otherwise, the algae would shade the bacteria and protect them from the ultraviolet rays. A large variety of methods for the isolation and purification has been described elsewhere (Kaushik, 1987).

For the precise qualitative distribution of soil algae, a known quantity of soil samples are to be inoculated in specific media and allowed to incubate in growth room or illuminated incubator. A periodic microscopic examination would reveal the various types of algae present.

REFERENCES

Allison, F.E. and Morris, H.J. 1929. Nitrogen fixation by blue-green algae. *Soil Sci.*, **71:** 221.

Bortels, H. 1940. Uber die Bedeutung des Molybdaens fuer Stickstoffbindene Nostoccaceen. *Arch. Mikrobiol.* **11:** 155.

Fogg, G.E. 1942. Studies on nitrogen fixation by blue-green algae. I. Nitrogen fixation by *Anabaena cylindrica* Lemm. *J. Exp. Biol.*, **19:** 78.

Friedmann, I. 1966. Microscopy of algal chromatophores. *Phycologia,* **6:** 29-36.

Godward, M.B.E. (ed.) 1966. *The chromosome of algae.* Edward Arnold, London.

Kaushik, B.D. 1987. *Laboratory Methods for Blue-green algae.* Assoc. Publ. Co., New Delhi, pp. 171.

Nygaard, G. 1977. On making fluid mounts of plankton algae. *Phycologia,* **16:** 351.

Parker, B.C. and Diboll, A.G. 1966. Alcian stains for histochemicals localization of acid and sulphated polysaccharides in algae. *Phycologia,* **6:** 37-46.

Pringsheim, E.G. 1914. Zur Physiologie der Schizophyceen. *Beitr. Biol. Pfl.* **12:** 49.

Sundberg, I. and Lembi, C.A. 1976. Phosphotungstic acid-chromic acid : a selective stain for algal plasma membranes. *J. Phycol.* **12:** 48-54.

APPENDIX

Aceto-carmine (Iron-aceto-carmine)

An acid dye. Stains and fixes. Differentiate in water.

Glacial acetic acid	45	ml
Distilled water	55	ml
Carmine (powdered)	excess	

Dissolve the carmine in the acetic acid, just bring to boil, cool, filter. A drop or two of 45% acetic acid saturated with iron acetate can be added. Too much iron precipitate the carmine.

Acid Fuchsin

An acid dye. A good general stain. Also for cytoplasm, chromatophores; may be used as counterstain for iodine green, methyl green, analine blue and metachite green. May be used in alcoholic or in aqueous solution.

Acid fuchsin	1	gm
Alcohol (50%) (or distilled water)	100	ml

Acid water - dye differentiator

Used as dye differentiator

Distilled water	100	ml
Hydrochloric acid (conc.)	0.5	ml

Alcian dyes

Colour the specimen 30-60 min in 0.5% aqueous alcian blue (AB) adjusted to pH 0.5 with N HCl. Wash 10 sec in distilled water adjusted to pH 0.5. Wash well in distilled water. Colour 30-60 min in 0.5% aqueous alcian yellow adjusted to pH 2.5 with N HCl. Wash in water. Dehydrate, clear and mount. Alcian dyes may be used successfully to differentiate non-sulphated polysaccharides in blue-green algae (*Nostoc*).

Alum carmine

Ammonium alum	4	gm
Carmine	1	gm
Water	100	ml

Boil the aqueous solution of the alum with the carmine for 20 min. Filter and allow to cool. Stain for about 12 hrs and the material will not be overstained.

Canada Balsam

The dried resin may be dissolved either in xylol or benzyl alcohol. The latter hardens mor quickly and is generally better for microscopic work. Now-a-days dissolved canada balsam is also available and can be employed.

Chloro iodide zinc - A chemical reagent

Zinc chloride	20	gms
Water	8.5	ml
Dissolve and cool		
Potassium iodide	1.0	gm
Iodine	0.5	gm
Water	20	ml

Add drop by drop to the first solution, until a precipitate of iodine forms, which does not disappear on shaking. This takes about 1.5 ml. Gives a violet colour to cellulose.

Euparal - mounting media

Euparal (colourless)	for mounting
Euparal (vert.)	for sealing

These two to be used, one for mounting and the other for sealing, the advantage here is that xylol is not required before mounting as in the case of balsam. Material can be mounted direct from 95% alcohol.

Gention violet stain

Gention violet	1	gm
Alcohol 95%	15	ml
Water	80	ml
Aniline oil	3	ml

A 1% solution in water without aniline keeps better. Gention violet may also be made up as a 1% solution in clove oil, or equal parts of absolute alcohol and clove oil.

This is a valuable stain for anatomical and cytological work.

Gram's Iodine

Iodine	1	gm
Potassium Iodide	2	gm
Water	100	ml

Lugol's Iodine solution

Iodine	1	gm
Pot. Iodide	2	gm
Water	300	ml

Meyer's albumin - fixative for sections

White of an egg	50	ml
Glycerine	50	ml
Sodium salicylate	1	gm

Separate the white from the yoke of the egg; dissolve the sodium salicylate in a very little water and add the other ingredients. The whole is mixed and then filtered.

Ind. Phyco. Rev. 1993.
2: 139-152.

PHYLOGENY AND INTERRELATIONSHIP

S.G. BHARATHI
Phycology Section, Department of Botany, Karnataka University, Dharwad - 580003.

Cyanophyceae (Sachs, 1874), also presently called the Cyanobacteria, are unique distinct group of algae belonging to the Protophyta. They show little differentiation in cell structure. The pigments are peripheral and glycogen is found as the food reserve. They also have a well defined sheath and there is complete absence of sexual reproduction and flagellated reproductive structures. All the above mentioned features show that this group is archaic and has not gone through much change throughout their period of existence.

Cyanophyceae had an independent origin from other classes of algae. They have a very low organisation. Their nucleus is not differentiated. They divide by ingrowth of septa and the presence of phycoerythrin is an indication of their adaptability. The tactic movements of the earth must have caused the irritation to induce unidirectional division leading to thread like forms. The complete absence of motile structures goes to show that these must have originated from unicellular forms that had no flagella. The simplest forms are found in the Chroococcales, where both coccoid and palmelloid conditions are present, and these appear to be parallel with the primitive motile condition present in the algae. The predominance of terrestrial forms in the blue-green algae (BGA) particularly in the Chroococcales suggests a terrestrial origin for the entire group.

BGA are broadly divided into two tribes (i) Coccogoneae with unicellular reproductive bodies and (ii) Hormogoneae reproducing by few celled short filaments-trichomes (Hormogones). Naegeli (1849), Hansgirg (1888) studied the Chroococcales and Thuret (1875), Bornet and Flahault (1885, 86, 88) and Gomont (1892) the Hormogonales separately. They divided the Hormogonales into Trichophoreae (forms with hair) and Psilonemeae (forms without hair), both these subtribes together have the following five families (i) Oscillatoriaceae (ii) Nostocaceae (iii) Scytonemataceae (iv) Stigonemaceae or Sirosiphonaceae and (v) Rivulariaceae. One more family Chamaesiphonaceae was created by Borzi (1878) to include genera like *Chamaesiphon* and *Clastidium*.

Both branched and unbranched forms are included in the filamentous forms. The branching is of two types-true and false branching. Based on the branching, the Hormogonales are divided into orders Nostocales and the Stigonematales (Geitler 1925, Fritsch 1942). The heterocysts, their presence and absence has been used as a criteria for separation of some genera. The formation of either the heterocyst or akinete may be minor genetic phenomenon and as such these are not completely dependable characters. The application of microscopic studies has been suggested as a means to clear doubts on taxonomic problems.

The division on the basis of presence or absence of heterocysts had many anomalies. Many beaded filaments characteristic of Nostocaceae do not develop heterocysts. The presence of heterocysts distinguishes Rivulariaceae but *Anabaena* also tapers as reported by Bornet & Flahault (1886-88). They limited the family Nostocaceae to eight genera. *Anabaenopsis, Richelia* and *Pseudanabaena* were included later. According to Fritsch (1945) *Pseudanabaena* is more closely related to Nostocaceae as a nonheterocystous form than to Oscillatoriaceae. Species of *Anabaena* sometimes do not show formation of heterocysts even in the same species. Another doubtful species is *Isocystis* which has been placed at the end of Nostocaceae by Bornet & Flahault (1880), in Oscillatoriaceae by Geitler (1932) and then again in Nostocaceae (Geitler 1942). Fritsch (1945) however keeps it in Nostocaceae on account of the presence of spores in chains.

Fritsch & Rich (1929) created a new genus *Raphidiopsis curvata* based on the bristle like acicular ends with absence of heterocysts, and less frequent akinetes. Skuja (1937), and Singh (1942) described two more similar species. All these show features common to Nostocaceae (*Anabaenopsis raciborskii & Aphanizomenon flos-aquae)*. Drouet (1938) described doubtful species with heterocysts, but the present one has heterocysts, which are intercalary. According to Prouse there is no justification in maintaining *Rhaphidiopsis, Aphanizomenon* and *Anabaena* as separate genera.

Lemmermann (1910) created Microchaetaceae to include forms with sheath like *Microchaetae, Aulosira* and *Desmonema*. Geitler (1932) removed *Desmonema* to Scytonemataceae and included *Hormothamnion* in the Microchaetaceae. Fritsch (1949) believes that *Aulosira* shows greater affinity to Nostocaceae than with Microchaetaceae in view of the distinction between base and apex and the presence of basal heterocyst. According to Bornet and Flahault (1880) *Aulosira and Hormothamnion* resemble *Nodularia* and all together should be placed in Nostocaceae and suggest the creation of two subfamilies Anabaenaceae and Aulosireae. The latter to include *Aulosira, Hormothamnion, Nodularia.* Elenkin (1938) also pointed a similar affinity by creating another family Nodulariaceae to include *Nodularia, Aulosira* and *Microchaete.*

Richelia with heterocysts at both ends is more related to Nostocaceae than Microchaetaceae. Fritsch (1942) however places it in Microchaetaceae.

Elenkin (1938) divides the genera with nostocaceous affinity into three families, Aphanizomeanaceae, Nostocaceae and Anabaenaceae and also divides *Nostoc* into a number of genera like *Amorphonostoc, Sphaeronostoc, Stratonostoc* and *Nematonostoc.*

Drouet (1951) places *Microchaetae* and *Aulosira* in Scytonemataceae and places *Hormothamnion* in Nostocaceae. *Raphidiopsis* like *Isocystis* is more related to Nostocaceae than to the other families.

Randhawa (1936) created three species of *Anabaena* that had sheathed trichomes, Rao, C.B. (1937), Geitler (1942), Fritsch (1945) consider them as not distinct from *Anabaena*.

Some of the families in the Cyanophyceae are not so distinct, therefore one feels the necessity to unite them into one family (e.g. Nostochopsidaceae, Mastigocladaceae, Mastigocladopsidaceae and Stigonemataceae). Since these have intermediate characters, Borzi (1914-1919) included all these in one family Stigonemataceae and made a number of subfamilies in it.

The Hormogoneae have been divided on the basis of the presence or absence of hair by Bornet & Flahault (1886-1888) and Gomont (1892). By this method distinct genera like *Mastigocoleus, Brachytrichia* and *Rivularia* will have to be placed together which is not justified. (Fritsch 1944, Papenfuss 1955). The close relationship of genera like *Plectonema, Scytonema, Homoeothrix* and *Calothrix, Pseudanabaena* and *Anabaena* will be affected if heterocyst is taken as a major character for the division of the Hormogoneae. Elenkin (1934, 36) recognised the importance of the heterocyst and used it to distinguish the Heterocysteae and Aheterocysteae and in the latter he included one order the Oscillatoriales and fourteen other distinct families in the Heterocysteae.

Gomont (1892) first recognised *Arthrospira* and many other aseptate forms which have been chemically demostrated as having septa. (Garder 1917 and others). Geitler (1925, 32, 42) recognised only one genus *Spirulina* and split it into two subsections *Arthrospira* (Stizenb.) Geitler and *Euspirulina* Forti. Crow (1927) studying in greater detail found many unseptate forms in *Spirulina* and therefore, limited *Arthrospira* only to septate forms. But Fritsch (1945), Drouet (1961) find it convenient to retain both the genera separately. Elenkin (1925) first placed *Spirulina* in Chroococcaceae as a unicellular form but later (1936, 49) transferred it back to Oscillatoriaceae.

In recent years attempts have been made to use other criteria like reproduction for classifying the BGA. Mode of reproduction in the Chamaesiphonales has been worked out in detail by Geitler (1925 a & b). The algae classed as Chamaesiphonales have numerous additions made during the last twenty years. They include two distinct series of plants viz. a group of uni or multicellular epiphytes-which may be called Chamaesiphonales and a second striking series of heterotrichous epiphytes or lithophytes, for them the name Pleurocapsales is suggested.

Chamaesiphonales is closely allied to Chroococcales and probably represent an offshoot from the same line of evolution. Affinities of Pleurocapsales is difficult to determine. The Chroococcales are traditionally divided into two familes

Chroococcaceae and Entophysalidaceae. The family Entophysalidaceae was raised to the order Entophysalidales by Geitler (1925). In 1932, he abolished this order and along with Chlorogloeaceae included it in the Chroococcales.

Elenkin (1936, 38) separated *Johannesbaptistia* and *Tubiella* into Tubiellaceae under the Tubiellales, the rest of the members were, however, retained in Entophysalidaceae, which, because of their great resemblance to coccoid forms the family was retained in Chlorococcales as representing the development of a filamentous tendency.

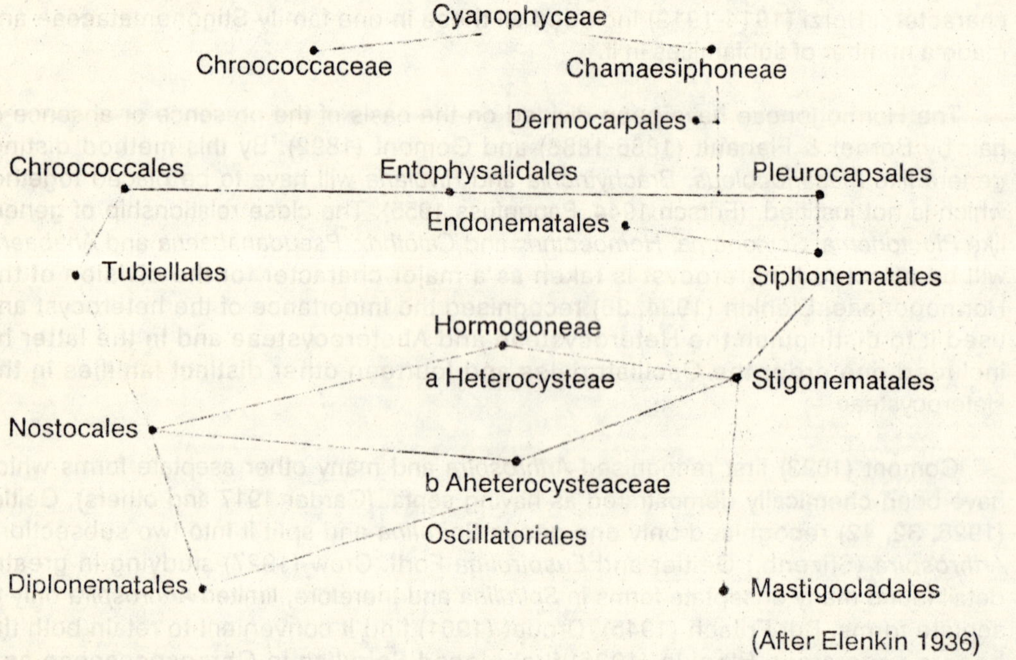

(After Elenkin 1936)

Drouet and Daily (1956) reduced all the genera to one genus *Entophysalis* including the same in Chamaesiphonaceae. *Johannesbaptistia* however was included in the Chroococcaceae. The Chamaesiphonales are kept distinct. The third order Pleurocapsales shows a heterotrichous condition without heterocysts. Some authors include all the three orders in the general term coccoid forms as against Hormogonales. Padmaja (1972) observes that coccoid BGA are ordinarily demarcated on dimensions of cells, colony, spores etc. Drouet & Daily (1956) used the habits for separating the species. Their classification is essentially based on the type method and not on the descriptions and illustrations given by the authors. Members of the Chroococaceae show considerable variation as a result the separation of genera and families is not constant, therefore Elenkin (1923) recognised fourteen genera and reduced others to subgenera and later in 1934 divided them into ten families with eight genera. Drouet & Daily (1952, 1956) studying on all the available type materials recognised only six genera. Fritsch (1942, 45) on the basis of differentiation of base and apex separated the forms and based on this Desikachary (1955) recognised seventeen genera. Considering the later revision by Drouet on Oscillatoriaceae (Drouet 1968) and Nostocaceae (Drouet 1973, 1978), as inconsistent for identification Padmaja and Desikachary (1967) called for a conservative approach towards modification of existing taxonomy.

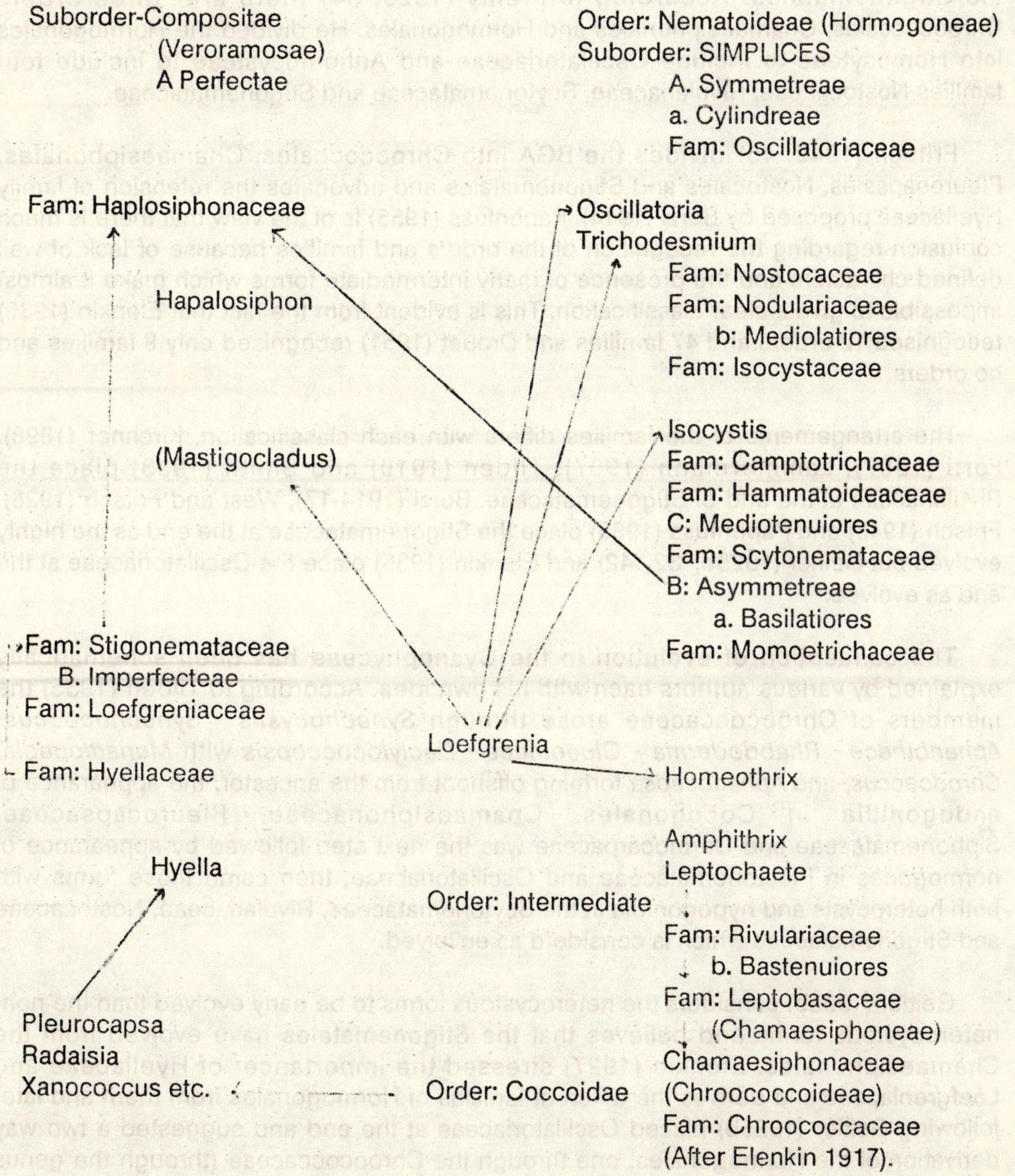

Borzi's (1978, 1895, 1914 to 17) studies on the Stigonemaceae clearly recognises two groups in it, the Diplonemaceae and Stigonemaceae on the basis of the division. In Stigonemae he included four groups Lorielleae, Capsosireae, Eustigonemeae and Nostochopsideae. Geitler (1925 b) elevated these to distinct families and in 1942 he recognised four orders Chroococcales, Dermocarpales, Pleurocapsales, Hormogonales and in the last order he made 13 families and dropped two orders the Nostocales and the Stigonematales. According to Fremy (1929, 34) there are three orders Chroococcales, Chamaesiphonales and Hormogonales. He divided the Hormogonales into Homocyteae to include Oscillatoriaceae and Anhomocysteae to include four families Nostocaceae, Rivulariaceae, Scytonemataceae and Stigonemataceae.

Fritsch (1942, 45) divides the BGA into Chroococcales, Chamaesiphonales, Pleurocapsales, Nostocales and Stigonematales and advocates the retension of family Hyellaceae proposed by Borzi (1914). Papenfuss (1955) is of the view that there is much confusion regarding the recognition of the orders and families because of lack of well defined characters and the presence of many intermediate forms which make it almost impossible to give a clear classification. This is evident from the fact that Elenkin (1936) recognised 12 orders and 47 families and Drouet (1951) recognised only 8 families and no orders.

The arrangements of the families differs with each classification, Kirchner (1898), Forti (1907), Lammermann (1907), Tilden (1910) and Smith (1933) place the Rivulariaceae at the end of Stigonemataceae. Borzi (1914-17), West and Fritsch (1928), Fritsch (1945) and Papenfuss (1955) place the Stigonemataceae at the end as the highly evolved but Geitler (1925b, '32, '42) and Elenkin (1936) place the Oscillatoriaceae at the end as evolved.

The conception of evolution in the Cyanophyceae has been schematically explained by various authors each with his own idea. According to Tilden (1935) the members of Chroococcaceae arose through *Synechocystis - Synechococcus- Aphanothece - Rhabdoderma - Gloeothece - Dactylococcopsis* with *Merismopedia, Chroococcus,* and *Aphanocapsa* forming offshoot from the ancestor, the appearance of endogonidia in Cocogonales, Chamaesiphonaceae, Pleurocapsaceae, Siphonemataceae and Dermocarpaceae was the next step followed by appearance of hormogones in Plectonemataceae and Oscillatoriaceae, then came those forms with both heterocysts and hypogonidia in the Scytonemataceae, Rivulariaceae, Nostocaceae and Stigonemataceae which is considerd as envolved.

Geitler (1925b) considers the heterocystous forms to be early evolved than the non-heterocystous forms and believes that the Stigonematales have evolved from the Chamaesiphonales. Elenkin (1927) stressed the importance of Hyellaceae and Loefgreniaceae and derives the different families of Hormogonales from them and later following Geitler (1925b) placed Oscillatoriaceae at the end and suggested a two way derivation of the Hormogonales, one through the Chroococcaceae (through the genus *Tubiella*) and the second way through *Siphonema.*

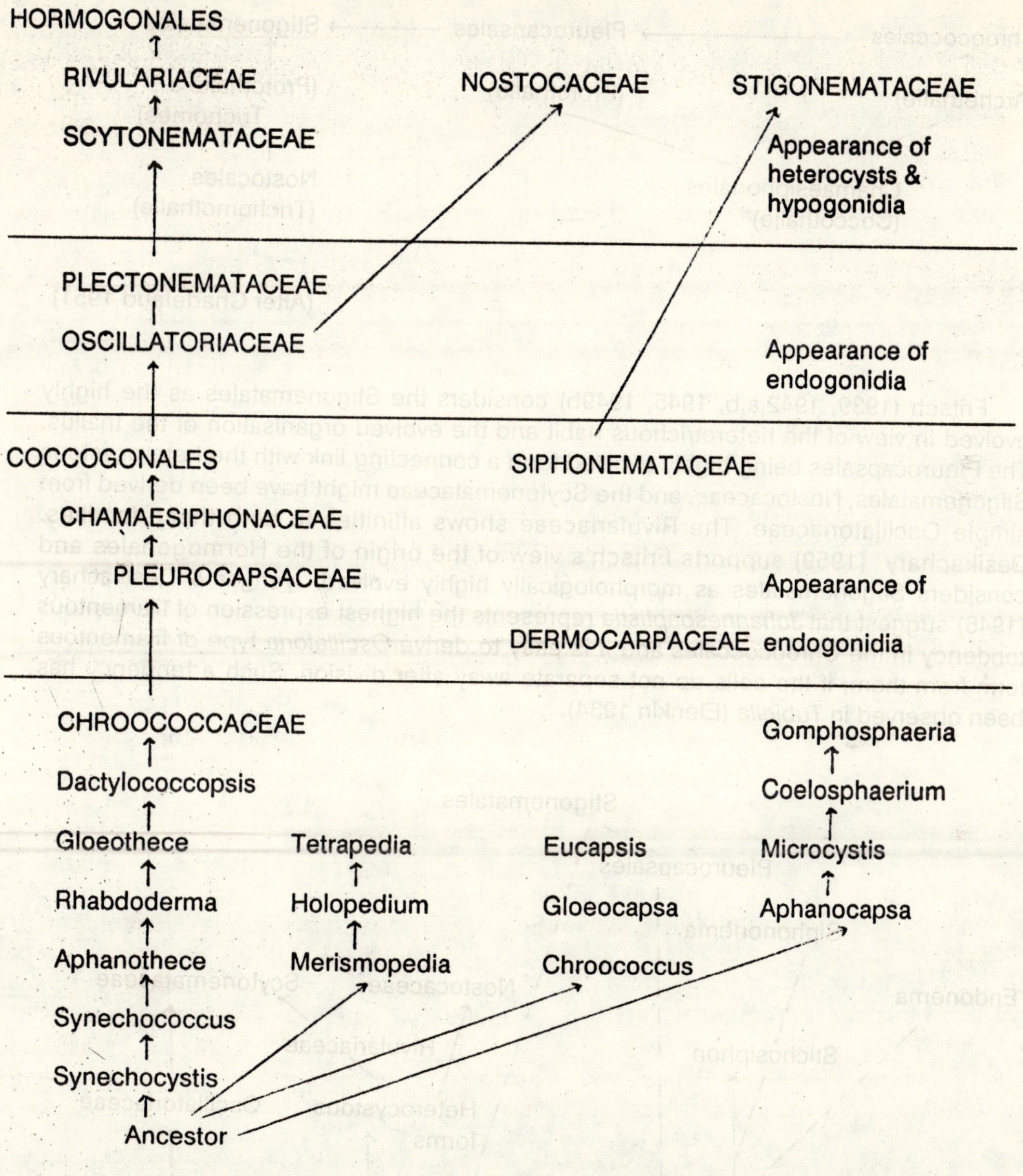

(Tilden 1935)

Chadefaud (1951) described the chroococcaceous genera as having an archthallus reproducing by non-motile spores and prothallus forms developed with polystichous and monostichous conditions and reproducing by endospores and hormospores which led to forms in th Pleurocapsales and the Stigonemales from which later evolved the Nostocales from which the Hormoganales (Trichonothallus) probably were derived from both the Chroococcales and the Pleurocapsales.

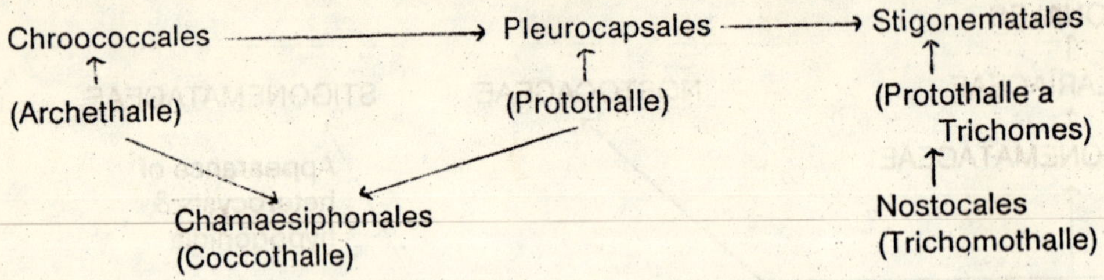

(After Chadefaud 1951)

Fritsch (1939, 1942,a,b, 1945, 1949b) considers the Stigonematales as the highly evolved in view of the heterotrichous habit and the evolved organisation of the thallus. The Pleurocapsales being highly evolved form a connecting link with the heterocystous Stigonematales, Nostocaceae, and the Scytonemataceae might have been derived from simple Oscillatoriaceae. The Rivulariaceae shows affinities to both these families. Desikachary (1959) supports Fritsch's view of the origin of the Hormogonales and considers Stigonematales as morphologically highly evolved. Iyengar & Desikachary (1946) suggest that *Johannesbaptistia* represents the highest expression of filamentous tendency in the Chroococcales and it is easy to derive *Oscillatoria* type of filamentous form from them; if the cells do not separate away after division. Such a tendency has been observed in *Tubiella* (Elenkin 1934).

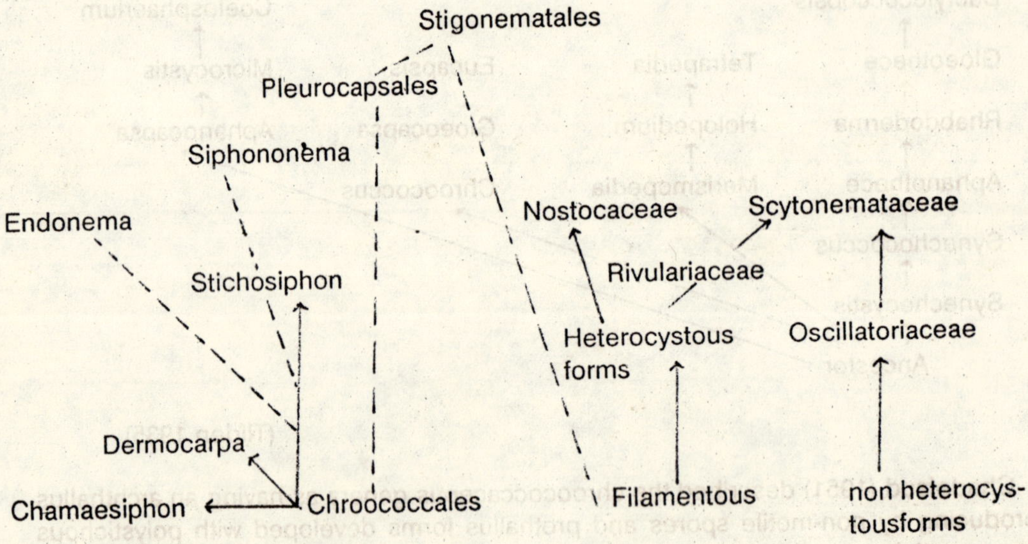

(Fritsch 1942)

Desikachary (1959) feels that both Nostocales and Stigonematales need to be separated in view of the branching which is characteristic feature of the BGA. Geitler (1932) suggests that the filamentous tendency seen in the Entophysalideceae might be ancestral to the Pleurocapsales which give rise to erect branches. Mitra (1951) reports a new species *Chlorogloea fritschi*, the endospores of which on germination produce a uniseriate filament which by further division and maturation gives rise to *Microcystis* like alga. This clearly supports the contention of Geitler about the derivation of Pleurocapsales from the Entophysalidaceae.

Scytonemataceae can be derived from *Lyngbya* like ancestor through *Plectonema* by developing a heterocyst or from forms like *Nodularia* and *Aulosira*. *Rivularia* also shows affinities to more than one family. Bornet & Flahault (1886) include the following ten genera *Polythrix, Isactis, Sacconema, Brachytrichia, Leptochaete, Amphithrix, Gloeotrichia, Dichothrix, Rivularia* and *Calothrix* in Rivulariaceae. Krichner (1898) and Forti (1907) added *Loefgrenia* and did not recognise *Gloeotrichia* as a distinct genus. Later *Hammatoidea* and *Tapinothrix* were added by Lemmermann (1910) and Geitler (1925). Fritsch & Rich (1924) added one more genus *Raphidiopsis*.

Elenkin (1936, 49) separated *Leptochaete, Homoeothrix* and *Amphithrix* to Homoeotrichaceae and *Loefgrenia* to Loefgreniaceae. These two families are placed under Oscillatoriales as a homocystous family. The Rivulariaceae included *Calothrix, Rivularia, Gloeotrichia*. Poljansky (1928) included *Dichothrix, Isactis* and *Sacconema* in the genus *Calothrix* and divided it into three sections Eucalothrix, Dichothrix and Isactis. Desikachary (1959) places *Homoeothrix, Calothrix, Dichothrix, Rivularia, Gloeotrichia* and *Leptochaete* under Rivulariaceae, *Brachytrichia* in Mastigocladaceae and *Microchaete* by the development of apical hair or might be evolved from Scytonemataceae through a form like *Tildania*. The simplest Stigonemataceae can be derived from simple Nostocales by the appearance of true branching.

Fritsch (1945) suggests the retension of *Mastigocladus* in the Stigonemataceae and separation of *Brachytrichia* into a separate family on the basis of true branching. Desikachary (1959) however suggest the retension of both *Mastigocladus* and *Brachytrichia* in Stigonemataceae in view of the obliqe division of a single parent cell and also by the reverse 'v' shaped branching in a distinct family Mastigocladaceae in the order Stigonematales.

Borzi (1914) suggests the retension of Diplonemataceae in the Stigonematales and Desikachary (1959) supports it in view of the true branching although there is false branching which is also present in varying degree in *Hapalosiphon* sp.

The following is the synoptic arrangement of the families and orders in the group (Desikachary 1959).

Phylum *CYANOPHYTA* Smith (1938)
Class Cyanophyceae Sachs (1874)

Order *CHROOCOCCALES* Wettstein 1924
Family i Chroococcaceae Nageli 1849
Family ii Entophysalidaceae Geitler 1925b
* Order *CHAMAESIPHONALES* Wettstein 1924
Family i Cyanidiaceae Geitler 1933
Family ii Chamaesiphonaceae Borzi 1882
Order *PLEUROCAPSALES* Geitler 1925b
Family i Pleurocapsaceae Geitler 1925b
Family ii Hyallaceae Borzi 1914
Order *NOSTOCALES* Geitler 1925b
Family i Oscillatoriaceae (S.F. Gray) Dumont ex Kirch. 1898.
Family ii Gomontiellaceae Elenkin 1936
Family iii Nostocaceae Kutz 1943
Family iv Scytonemataceae Raben ex. Born et Flah (1887)
Family v Microchaetaceae Lemm. 1907.
Family vi Rivulariaceae Raben 1868
Order *STIGONEMATALES* Geitler 1925b
Family i Capsosiraceae Geitler 1925b
Family ii Pulvinulariaceae Geitler 1925b
Family iii Nostochopsidaceae Geitler 1925b
Family iv Diplonemataceae (Borzi) Elenkin 1934
Family v Mastigocladaceae Geitler 1925b
Family vi Mastigocladopsidaceae Iyeng. et Desik. 1946.
Family vii Stigonemataceae Kirchn. 1898.

* included as third family Chamaesiphonaceae under Chroococcales (Desikachary 1982).

REFERENCES

Bornet, E. & Thuret, G. 1880. Notes on algologiquer, II. Paris.

Bornet, E. & Flahault, C. 1895. Tableau synoptiques des Nostocacees filamenteuses Heterocystees. *Mem. Sco. Nat. Sci. Nat. et Math, Cherbourg* 25: 195-223.

— 1886. Revision des Nostocacees heterocystees. I. *Ann. Sci. nat. Bot. Ser.* 7, 3: 323-381.

— 1888. Revision des Nostocacees hetercystees. IV. *Ibid.* 7: 177-262.

Borzi, A. 1878. Nachtrage zur Morphologie und Biologie der Nostocaceen. *Flora* 61: 465-471.

— 1895. Studi algologici, Fasc. 2: 257 pp. Palermo.

— 1914. Studi sulle Mixoficee -I. *Nuovo Giorn. Bot. ital, n.s.* **21**: 307-360.

— 1914. Studi sulle Mixoficee - II. *Ibid.* **23**: 559-588.

— 1917. Studi sulle Mixoficee - III *Ibid.* **24**: 17 et seq.

Crow, W.B. 1927a. *Crinalium*, A new genus of Cyanophyceae and its bearing on the morphology of the group. *Ann. Bot., London.* **41**: 161-165.

— 1927b. The generic characters of *Arthrospira* and *Spirolina* Turp. *Trans. Amer. micr. Soc.* **46**: 139-418.

Chadefaud, M. 1951. L' evolution morphologique et la systematique des Cyanophycees. 70 e. congr. de L' A.F.A.S. 4: 3-8.

Desikachary, T.V. & Mall, L.P. 1955. Occurrence of *Iyengariella tirupatiensis* Desi. at Pachmarhi Hills. *Curr. Sci.* **24**: 245-246.

Desikachary, T.V. 1959. *Cyanophyta.* I.A.R.I., New Delhi. pp. 1-686.

Desikachary, T.V. 1982. Status of classical taxonomy. In: (eds. Carr,N.G. and Whitton,B.A.) *Biology of blue-green algae.* Blackwell Scientific Publications, London.

Drouet, F. 1938. Myxophyceae of the Yale North India Expedition. Coll. by G.E. Hutchinson. *Trans. Amer. micr. Soc.,* 57(2).

— 1951. Cyanophyta. In: *Manual of Phycology* (ed. G.M. Smith). Chronica Botanica. Waltham. Mass.: 159-166.

Drouet, F. 1968. Revision of the classification of the Oscillatoriacees. *Monogr. Acad. nat. Sci. Philadelphia,* **16**: 1-341.

Drouet, F. 1973. Revision of Nostocaceae with cylindrical trichome. Hafner Press. N.Y. 192 pp.

Drouet, F. 1978. Revision of Nostocaceae with constricted trichomes. *Beihfte zur Nova Hedwigia Heft* **47**: 258 pp.

Drouet, F. & Daily, W. 1952. A synopsis of the coccoid Myxophyceae. *Butler Uni. studies. Bot.* **10**: 120-223.

— 1956. Revision of the coccoid Myxophyceae. *Ibid.* **12**: 1-218.

Elenkin, A. 1933. Uber neue Familien der cyanophyceen aus der Gruppe der stereometreae Elenkin (untergruppe Excavate Elenk) der ordnung der chroococcales. Geitler (1925). *Act Inst. Bot Acad. sci. U.R.S.S.* ber 2, fasc. **1: 17-21, 27-34.**

Elenkin, A. 1936. Monographie algarum cyanophycearum aquidulcum et terrestrium infinibus U.R.S.S., Moscow, Leningard.

--- 1938. *Ibid.* Pars specialis (Systematica) Fasc. I., 984 pp. Acad. naute, U.R.S.S. Moscow, Leningard.

---- 1949. Monographie algarum Cyanophycearum acquidulcium et terrestrium infinibus U.R.S.S. inventarum. Vol. I. 675 pp. Pars specialis (Systematica) Fasc. II, 985-1908 pp. Acad. Nauk. U.R.S.S. Moscow. Leningard.

Forti, A. 1907. Myxophyceae in De Toni. Sylloge *Algarum* **5:** 761 pp. Padova.

Fremy, P. 1929. Les Myxophycees de l'Afrique equatorale francaise. *Arch. Bot. caen,* **3:** Memoire no.2.

Fremy, P. 1934, Les Cyanophyceae des cpes d'Europe. *Mem-Soc. nat Math de cherboury,* **41:** 1-236.

Fritsch, F.E. 1942. The interrelations and classification of the Myxophyceae (Cyanophyceae). *New Phytol.* **41:** 134-148.

— 1945. The structure and reproduction of Algae, Cambridge. Vol. 2. 939. pp.

— 1949b. The lines of algal advance. *Biol. Rev.* **24:** 94-124.

Geitler, L. 1925a. Uber neue oder wenig bekannte interessante Cyanophyceen und der Gruppe der Chamaesiphoneae. *Arch. Protistenk.* **51:** 321-360.

--- 1925b. Synoptische Darstellung der Cyanophyceen in morphologische und Systematisher Hinsicht. *Bein. bat. cbl.* II. **41:** 163-294.

— 1932. Cyanophyceae in Rabenhorst's Kryptogamenflora, *Leipzig.* **14:** 196 pp.

— 1942. Schizophyceae in Engler-Prantl, Die Naturalichen Pflangenfam lien. IInd Ed., Ia.

Gomont, M. 1892a. Monographie des Oscillarees I. *Ann. Sci.nat. Bot.* ser. 7, **15:** 263-369.

— 1892b. Monographie des Oscillarees II. *Ibid.* **16:** 91-264.

Hansgirg, A. 1888b. Synopsis generumsubgenerumque Myxophycearum (Cyanophycearum) hycusque congnitorum cum descriptione generis nov. Dactylococcopsis. *Notarisia* **3: 548-590.**

Kirchner, O. 1900. Schizophyceae in Engler & Prantl, Die Naturlichen Pflangenfem, I.la. 45-92.

Korde, K.B. 1972. Ancient blue green algae. Taxon. & Biol. of BGA. pp. 161-166.

Lemmermann, E. 1910. Algae I. in Kryptogamenflora der Mark Brandenburg, 1256, Leipzig. 3.

--- 1904. Uber die von Herrn Dr. Walter volz auf seiner weltreise gesammalten Susswsseralgen. *Abh. Nat. ver. Bremen* **18:** 143-174.

Mitra, A.K. 1950. Two new algae from Indian soils. *Ann. Bot. Lond.* n.s. **14:** 457-464.

Naegeli, C. 1849. Gattungen einzelliger Algen, Physiologisch und systematisch bearbeitet, Neue Denkschriften Allg. Schwiz. Natur. Ges. 10: 139 pp. Zurich.

Padmaja, T.D. 1972. Studies on coccoid BGA. II. *Symp. Taxonomy & Biology of Blue-green algae* (ed. T.V. Desikachary). : 75-127.

Padmaja, T.D. and Desikachary, T.V. 1967. Trends in the taxonomy of algae. *Bull. Natn. Inst. Sci.* **34**: 338-364.

Papenfuss, G.F. 1955. Classification of the Algae, In: *A Century of progress in Natural Sciences* (1893-1953). Calif. Acad.Sci.San Francisco. 115-224.

Poljansky, V. 1928a. Uber das gegesietige verhaltnis der Families Rivulariaceae (Menegh.) Elenk. und Tildeniaceae. Kossinsk., auf grund einer kristischen einiger *Calothrix*-arten. J. Congr. U.R.S.S. Bot. Leningrad 158-159.

Radhawa, M.S. 1936b. Genus *Anabaenothrix* and parallelism in freshwater algae. *Proc. Ind. Acad. Sci.* B. **3**(5): 401-409.

Rao, C.B. 1937. A new species of *Anabaena.* (*A.ambigua* sp. nov.). *Proc. Ind. Acad. Sci.*B. **2**(6): 535-538.

Sachs, J. 1874. Lehrbuch der Botanik. 4th Edn. XVI. 928 pp. Leipzig. W. Engelmann.

Skuja, H. 1937. Algae in Handel-Mazzetti, H., Symbolae Sinicae, I. Vienna.

Singh,V.P. 1941. On a collection of algae from Chamba State (Punjab) I. *Proc. Ind. Acad. Sci.* B. **14:** 250-255.

Stizenberger, E. 1860. Dr. L.Rabenhorst's Algen Sachsens etc. systematisch geordnet mit zugrundlegung eines neuens systems, Dresden.

Smith, G.M. 1950. Freshwater algae of the United States of America. McGraw Hill, New York. 2nd Ed. 719 pp.

Thuret, G. 1875. Essai de classification des Nostochinees. *Ann. nat. Bot.* ser. 6, 1: 372-382.

Tilden, J.E. 1910. Myxophyceae in Minnesota Algae, I. Minneapolis, 550 pp. 1935. Algae and their life relations, 550 pp. Minneapolis.

West, G.S. & Fritsch, F.E. 1927. A treatise in British Freshwater algae. Cambridge.

APPENDIX

Research Laboratories in India working on blue-greens

C.A.S. in Botany, University of Madras, Madras - 600 025
C.A.S. in Botany, Banaras Hindu University, Varanasi, U.P.
National Facility for Blue-green algae, Indian Agricultural Research Institute, New Delhi.
DBT Centre for Blue-green algal Biofertilizers, Madurai Kamaraj University, Madurai.
DBT Centre for Marine cyanobacteria, Bharathidasan University, Trichy.
Bhabha Atomic Research Centre, Bombay.
Central Rice Research Institute, Cuttack, Orissa.
Central Food Technological Research Institute, Mysore.
Tamil Nadu Agricultural University, Coimbatore.
University of Agricultural Sciences, GKVK Campus, Bangalore.
North Eastern Hill University, Shillong.
Department of Life Sciences, Central University, Hyderabad.
Department of Botany, Allahabad University, Allahabad.
Department of Botany, Punjab University, Chandigarh.
Department of Botany, Berhampur University, Berhampur, Orissa.
Institute of Botany, Nagpur.
Department of Botany, Andhra University, Waltair.
Murugappa Chettiar Research Centre, Taramani, Madras.
Department of Botany, M.S. University, Udaipur.
Punjabi University, Patiala.
Department of Botany, Gulbarga University, Gulbarga.
Department of Botany, Karnataka University, Dharwad.
Department of Botany, Marathwada University, Aurangabad.
Department of Botany, Utkal University, Utkal, Orissa.
Jawaharlal Nehru University, New Delhi.
Department of Microbiology, Barkhatullah University, Bhopal.
Department of Botany, Bihar University, Muzaffarpur, Bihar.
Phycological Laboratory, NBRI, Lucknow, U.P.